Photod
and Mea

Photodetection and Measurement

Maximizing Performance in Optical Systems

Mark Johnson

McGraw-Hill
New York   Chicago   San Francisco   Lisbon   London   Madrid
Mexico City   Milan   New Delhi   San Juan   Seoul
Singapore   Sydney   Toronto

The McGraw·Hill Companies

Library of Congress Cataloging-in-Publication Data

Johnson, Mark, date.
 Photodetection and measurement: maximizing performance in optical
 systems / Mark Johnson.
    p.   cm.
 Includes bibliographical references and index.
 ISBN 0-07-140944-0 (alk. paper)
 1. Photodetectors.   2. Optoelectronic devices.   3. Light—Measurement.
 I. Title.

 TK8360.O67J64   2003
 621.36—dc21   2003046318

 3 4 5 6 7 8 9 0   IBT/IBT   0 9

ISBN 0-07-140944-0

*The sponsoring editor for this book was Stephen S. Chapman and the production
supervisor was Sherri Souffrance. It was set in Century Schoolbook by SNP
Best-set Typesetter Ltd., Hong Kong.*

Printed and bound by IBT Global.

 This book was printed on recycled, acid-free paper containing a minimum of 50% recycled, de-inked fiber.

McGraw-Hill books are available at special quantity discounts to use as premiums
and sales promotions, or for use in corporate training programs. For more
information, please write to the Director of Special Sales, Professional Publishing,
McGraw-Hill, Two Penn Plaza, New York, NY 10121-2298. Or contact your local
bookstore.

Contents

Preface

The earliest I can recall an interest in optical detection was sometime around 1960. It was a project for an optically controlled model boat. All I can remember was that it used a handheld flashlight as transmitter and a light-dependent resistor on the model as receiver. Some rudimentary relay electronics taken from radio control provided the logic and memory to turn the boat on command, left, right, or straight ahead. I seem to remember it required some dexterity in flashing the light for about the right pulse-width to make it all work. I was desperate to make this, but with electronics knowledge limited to wiring batteries and bulbs, it was a hopeless endeavor.

Three years later, at school, we received some basic introduction to electronics, and it dawned on me that this would help with the old boat project. However, most of the time seemed to be spent with hole-cutters mounting valve sockets in aluminum chassis, winding component leads irretrievably around tag-strip connections, or in clearing the smoke from not-quite-correctly wired circuits. This wasn't what I had in mind for electronics, but it did at least engender some respect for high voltages, Ohm's law, and the ultimate thermal limitations of components. At about this time transistors became widely available, and I dropped thermionics almost for good.

When I started my research for a Ph.D. at London University, my colleagues and I needed to perform optical measurements, basic spectroscopy, loss measurements in planar optical waveguides, and optical fiber sensors demonstrations. It was usually possible to scrounge an old unmarked photodiode, and everyone seemed to point you to a high-performance electrometer from Keithley Instruments (analog of course). This was the pukka kit to use! However, no one seemed to have any idea how to use this beautiful voltmeter/ammeter. The technique seemed to be to connect the photodiode, spin the large rotary range selector until a signal was obtained, and see if it changed with your hand in front of the diode! I remember being incensed that no one knew the way to use it, or even whether the volts, amps or ohms scales would be best for my measurement.

That was thirty years ago, and since then I have had any number of similar experiences of "cluelessness." Generally I made the experiment work, and even had a few novel ideas along the way. At each impasse I tried to figure out what was needed to get it to work at all, and then to deliver the best performance possible. For almost every occasion, I realized years later how I *should* have done it to make it work much better.

In university labs there is usually a big community of "experts" who will tell you in broad terms how to solve some optoelectronic problem, or who explain in the mannerisms of an "old-soldier" how well they solved the identical problem years ago in some long-forgotten project whose details and project notes they can't recall, or who will "helpfully" say that you should be using a transimpedance amplifier or a Fourier transform or an avalanche photodiode, but when pinned down can't quite explain why, even less how actually to do it. So much voodoo is, for me, neither helpful nor encouraging. When all you want to do is solve the problem, it is much better if your guru can say "Do it this way, and it will work for these reasons." Almost as helpful is "I don't know, but let's go and find out."

Of course, it's not all voodoo. Several colleagues really have said "this is how you do it," and helped me greatly in the progress of many projects. One of my first was Bill McGarry at IBM's Thomas Watson Research Center, with his predilection for lots of gain, tamed with feedback, and back-of-the-envelope calculations of all the important issues. Robert Theobald of York Technology, with his feel for what kT/q really means, his vast store of issues convincingly thought through, and for a passion for making the "next one even better." Brian Elliott was also a fabulous inspiration, with his numerical rigor and dogged tenacity in understanding low-level electronic measurements and the physical processes underlying them, not just for electronics' sake but for the measurement or process under study. Then there were the companies, Ferranti, Mullard, Nexus, Philbrick, Analog-Devices, Burr-Brown, Zetex, and National Semiconductor, whose data-books and application notes I absorbed in a frenzy, and which sometimes changed the way I viewed the whole subject. And even the hobby stores, with their inspiring do-it-yourself projects, at once grossly overpriced on a cost-per-transistor level, and trivially cheap in didactic value.

Last, thanks to the Université Jean Monnet at St. Etienne in France, which gave me a month to lecture on the subject of optical detection, to try to find out what is important to the audience, and to think how to answer some of the difficult questions they posed. Without those questions, I don't suppose I would have ever gotten around to starting this book.

I know that this is not the most erudite work on the subject, and my level of understanding is a shadow of the authors whose books I read. Nevertheless, I seem to be able to see what might work, and to make it work in the limited sphere of laboratory optical measurements. Who is it for? The majority of physicists starting research work haven't done any electronics, at least not studied electronics in the sense of a real electronics course or spent ten years making his or her own HiFi and model aircraft. Professors will say that it doesn't matter, that the electronics department technicians will put together whatever you need. In practice they only do what you tell them, and, anyway, the little bits of electronics you need are so finely sprinkled throughout every working day that outsourcing is not efficient. You need to do it yourself. They even imply that electronics is somehow separate from physics, a kind of low-level support for your higher ideals. In my experience university research is littered with examples of good idea projects that were trashed because the first trials didn't work. They probably didn't work because the front end photodetector design was awful, the signal-to-noise ratio was a thousand times worse than it might have been, and it was connected to a computerized data acquisition system that didn't let you see how bad it was or how it could be fixed.

This book is for the physicist who has to get by in photo-measurement. It is designed to teach the new researcher working with modern optics the absolute basics of photodetection and the electronics that is so important to it, so that he can do good measurements, and spend more time thinking about the experiment. It is supposed to be practical, hands on, and tell you how you might really make it work. If it says an FET can make a good chopper, the next question has to be how? A useful text should say how.

Throughout the text are little experiments chosen to illustrate a point: TRY IT! They don't take much time (some are really basic) but they hopefully give a much better feel for what to expect in practice. Also in the text you will find occasional nonmainstream topics. These are aimed as an inspiration to look more widely at photodetection, modulation, coding, and even the mathematics of these processes, because the techniques that have been developed are often so elegant, and a few elegant ideas go a long way. If you look carefully at even the simplest of effects, there is always so much to see. Quick solutions to opto-electronic problems should leave you more time to look. When I apply the principles described here to optoelectronic measurements, they *almost always* eventually work. I hope that they work for you too.

Mark Johnson

Photodetection
and Measurement

Photodetection Basics

1.1 Introduction

The junction photodiode that is the focus of this book has been described in detail in many other books and publications. Here only a few basics are given, so that you can use the photodiode effectively in real circuits. A simple model is presented that allows the main characteristics and limitations of real components to be understood. The ability to correctly derive the polarity of a photodiode's output, guess at what level of output current to expect, and have a feel for how detection speed depends on the attached load is necessary. The model we begin with has little to do with the typical real component fabricated using modern processing techniques; it is a schematic silicon pn-junction diode.

1.2 Junction Diodes/Photodiodes and Photodetection

Figure 1.1a shows two separate blocks of silicon. Silicon has a chemical valency of four, indicating simplistically that each silicon atom has four electron bonds, which usually link it to neighboring atoms. However, the lower block has been doped with a low concentration (typically 10^{13} to 10^{18} foreign atoms per cm^3) of a five-valent element, such as arsenic or phosphorus. Because these dopants have one more valency than is needed to satisfy neighboring silicon atoms, they have a free electron to donate to the lattice and are therefore called *donor* atoms. The donor atoms are bound in the silicon crystal lattice, but their extra electron can be easily ionized by thermal energy at room temperature to contribute to electrical conductivity. The extra electrons then in the conduction band are effectively free to travel throughout the bulk material. Because of the dominant presence of negatively charged conducting species, this doped material is called *n-type*.

By contrast, the upper block has been doped with an element such as boron,

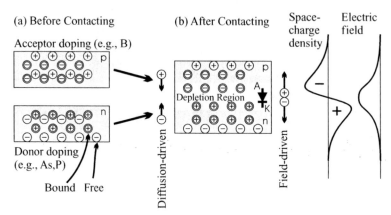

Figure 1.1 The pn-junction. The presence of predominately different polarities of free carriers in the two contacted materials leads to asymmetrical conductivity, a rectifying action. Bound charges are indicated by a double circle and free charges by a single circle.

which exhibits a valency of three. Because boron lacks sufficient electrons to satisfy the four surrounding silicon atoms and tries to accept one from the surrounding material it is termed an *acceptor* atom. As with donors, the bound boron atoms can easily be ionized, effectively transferring the missing electron to its conduction band, giving conduction by positive charge carriers or *holes*. The doped material is then termed *p-type*. The electrical conductivity of the two materials depends on the concentration of ionized dopant atoms and hence on the temperature. Because the separation of the donor energy level from the conduction band and the acceptor level from the valence band in silicon is very small (energy difference ≈ 0.02 to 0.05 eV) at room temperature the majority of the dopant atoms are ionized.

If the two doped silicon blocks are forced into intimate contact (Fig. 1.1*b*), the free carriers try to travel across the junction, driven by the concentration gradient. Hence free electrons from the lower n-type material migrate into the p-type material, and free holes migrate from the upper p-type into the n-type material. This charge flow constitutes the diffusion current, which tends to reduce the nonequilibrium charge density. In the immediate vicinity of the physical junction, the free charge carriers intermingle and recombine. This leads to a thin region that is relatively depleted of free carriers and renders it more highly resistive. This is called the *depletion region*. Although the free carriers have combined, the charged bound donor and acceptor atoms remain, giving rise to a space charge, negative in p-type and positive in n-type material and a real electric field then exists between the n- and p-type materials.

If a voltage source were applied positively to the p-type material, free holes would tend to be driven by the total electric field into the depletion region and on to the n-type side. A current would flow. The junction is then termed *forward-biased*. If, however, a negative voltage were applied to the p-type

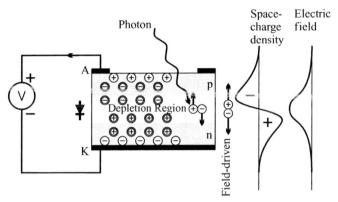

Figure 1.2 When a photon with energy greater than the material bandgap forms a hole-electron pair, a terminal voltage will be generated, positive at the p-type anode.

material, carriers would remain away from the depletion region and not contribute to conduction. The pn-junction is then called *reverse-biased*, and has very little current flow. Note that the diffusion currents driven by concentration gradients and the field currents driven by the electric field can have different directions. The conventional designation of the p-type contact is the anode (A); the n-type contact is the cathode (K).

This basic pn-junction diode model can also explain how a photodiode detector functions. Figure 1.2 shows the same diode depicted in Fig. 1.1 in schematic form, with its bound dopant atoms (double circled) and free charge carriers (single circled). A photon is incident on the junction; we assume that it has an energy greater than the material bandgap, which is sufficient to generate a hole-electron pair. If this happens in the depletion region, the two charges will be separated and accelerated by the electric field as shown. Electrons accelerate toward the positive space charge on the n-side, while holes move toward the p-type negative space charge. If the photodiode is not connected to an external circuit, the anode will become positively charged. If an external circuit is provided, current will flow from the anode to the cathode.

1.3 TRY IT! Junction Diode Sensitivity and Detection Polarity

The validity of this model can easily be tested. All diode rectifiers are to some extent photosensitive, including those not normally used for photodetection. If a glass encapsulated small signal diode such as the common 1N4148 is connected to a voltmeter as shown in Fig. 1.3 and illuminated strongly with light from a table lamp the anode will become positive with respect to the cathode. The efficiency of this photodiode is not high, as light access to the junction is almost occluded by the chip metallization. Nevertheless you should see a few tens of millivolts close to a bright desk lamp.

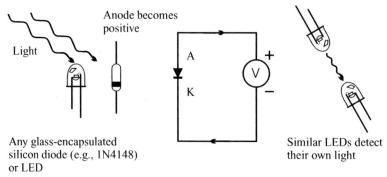

Figure 1.3 Any diode, even a silicon rectifier, can show photosensitivity if the light can get to the junction. LEDs generate higher open circuit voltages than the silicon diode when illuminated with light from a similar or shorter wavelength LED.

Rather more efficient are light emitting diodes (LEDs), having been designed to let light out of (and therefore into) the junction. Any common LED tested in this way will show a similar positive voltage on the anode. LEDs have the advantage of a higher open-circuit voltage over silicon diodes and photodiodes. This voltage gives an indication of the material's bandgap energy (E_g, see Table 1.1). Although with a silicon diode ($E_g \approx 1.1$ V) you might expect 0.5 V under an ordinary desk lamp, a red LED ($E_g \approx 2.1$ V) might manage more than 1 V, and a green LED ($E_g \approx 3.0$ V) almost 2 V. This is sufficient to directly drive the input stages of low voltage logic families such as 74 LVC, 74 AC, and 74 HC in simple detection circuits.

This works because the desk lamp emits a wide range of energies, sufficient to generate photoelectrons in all the diode materials mentioned. However, if the photon energy is insufficient, or the wavelength is too long, then a photocurrent will not be detected. My 470-nm blue LEDs generate negligible junction voltage under the desk lamp. Try illuminating different LEDs with light from a red source, such as a red-filtered desk lamp or a helium neon laser. You should detect a large photovoltage with the silicon diode, and perhaps the red LED, but not with the green LED. The bandgap in the green emitter is simply too large for red photons to excite photoelectrons. You can take this game even further if you have a good selection of LEDs. My 470-nm LED gets 1.4 V from a 660-nm red LED as detector but nothing reversing the illumination direction. Similarly the 470 nm generates 1.6 V from a 525-nm emitting green LED but nothing in return. These results were obtained by simply butting together the molded LED lenses, so the coupling efficiency is far from optimized. The above bandgap model suggests that LED detection is zero above the threshold wavelength and perfect below. In reality the response at shorter wavelengths is also limited by excessive material absorption. So they generally show a strongly peaked response only a few tens of nanometers wide, which can be very useful to reduce sensitivity to interfering optical sources. See Mims (2000) for a solar radiometer design using LEDs as selective photodetectors. Most LEDs are reasonable detectors of their own radiation, although the overlap of emission and detection spectra is not perfect. It can occasionally be useful to make bidirectional LED–LED optocouplers, even coupled with fat multimode fiber. Chapter 4 shows an application of an LED used simultaneously as emitter and detector of its *own* radiation.

For another detection demonstration, find a piece of silicon, connect it to the ground terminal of a laboratory oscilloscope and press a 10-MΩ probe against the top surface. Illuminate the contact point with a bright red LED modulated at 1 kHz. You should see a strong response on the scope display. This "cat's whisker" photodetector is about as simple a demonstration of photodetection as I can come up with! This isn't a semiconductor pn-junction diode, but a metal-semiconductor diode like a Schottky diode. It seems that almost any junction between dissimilar conducting materials will operate as a photodetector, including semiconductors, metals, electrolytes, and more fashionably organic semiconductors.

1.4 Real Fabrications

Although all pn-junction diodes are photosensitive, and a diode *can* be formed by pressing together two different semiconductor (or metal and semiconductor) materials in the manner of the first cat's whisker radio detectors or the previous TRY IT! demonstrations, for optimum and repeatable performance we usually turn to specially designed structures, those commercially produced. These are solid structures, formed, for example, by diffusing boron into an n-type silicon substrate as in Fig. 1.4 (similar to the Siemens BPW34). The diffusion is very shallow, typically only a few microns in total depth, and the pn-junction itself is thinner still. This structure is therefore modified with respect to the simple pn-junction, in that the diffusions are made in a high resistivity (intrinsic conduction only) material or additionally formed layer with a doping level as low as 10^{12} cm^{-3}, instead of the 10^{15} cm^{-3} of a normal pn-junction. This is the *pin*-junction photodiode, where "i" represents the thick, high-resistivity intrinsic region. Most photodetectors are fabricated in this way. The design gives a two order of magnitude increase in the width of the space-charge region. As photodetection occurs only if charge pairs are generated close to the high-field depleted region of the structure, this helps to increase efficiency and

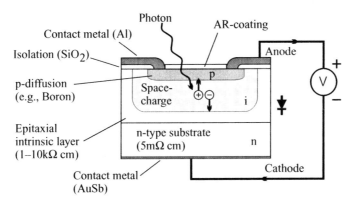

Figure 1.4   Most photodiodes are formed by diffusing dopants into epitaxially formed layers. The use of a low conductivity intrinsic layer leads to thickening of the space-charge region, lower capacitance, and improved sensitivity.

speed. Finally, additional high dopant concentration diffusions are performed to allow low-resistance "ohmic contact" connections to the top layer and substrate to be made for subsequent bonding of metal leads. We will return to this design in discussions of the sensitivity and wavelength characteristics of photodiodes.

1.5   Responsivity: What Current per Optical Watt?

Earlier we assumed that one photon generates one hole-electron pair. This is because detection in a photodiode, like the photoelectric effect at a vacuum-metal interface, is a quantum process. In an ideal case each photon with an energy greater than the semiconductor bandgap energy will generate precisely one hole-electron pair. Therefore, neglecting nonlinear multiphoton effects, the charge generated on photodetection above the bandgap is independent of photon energy. The photodiode user is generally most interested in the internal current (I_o) that is generated for each received watt (P_r) of incident light power. This is termed the responsivity (r) of the photodiode, with units of ampere per watt (A/W):

$$r = \frac{I_o}{P_r} \tag{1.1}$$

However, we will see later that the noise performance of the designer's circuitry is more a function of the arrival rate of photons at his detector and the total number of photoelectrons counted during his experiment. It is important to remember that detection is a quantum process, with the generation of discrete units of charge. As the energy of a photon (hc/λ) is inversely proportional to its wavelength, the number of photons arriving per second per watt of incident power is linearly proportional to the wavelength, and the responsivity of an ideal photodiode increases with wavelength. For this ideal case we can write:

$$r_{ideal}(\lambda) = \frac{I_o}{P_r} = \frac{q\lambda}{hc} \tag{1.2}$$

where λ = is the wavelength of the incident light in meters
$q = 1.602 \times 10^{-19}\,C$ is the charge on the electron
$h = 6.626 \times 10^{-34}\,Ws^2$ is Planck's constant
$c = 2.998 \times 10^8\,m/s$ is the speed of light in vacuum

For example, at a wavelength of $0.78\,\mu m$, the wavelength of the laser diode in a music CD player, $r_{ideal} = 0.63\,mA/mW$. So for a laser diode that emits $1\,mW$ or so, the change of units for responsivity into milliamperes per milliwatt is convenient and gives an immediate idea of the ($\approx mA$) photocurrent generated by its internal monitor diode. This is the responsivity for 100 percent quantum efficiency and for 100 percent extraction of the photocurrent.

Again with a view to obtaining expressions that can be used quickly in mental arithmetic, the wavelength can be expressed in microns to give:

$$r_{ideal} = 0.807\lambda_{\mu m} \quad \mathrm{mA/mW} \tag{1.3}$$

Hence, for a fixed incident power, the limiting performance of our detection systems is usually better at a longer wavelength. At longer wavelengths you simply have more photons arriving in the measurement period than at shorter wavelengths. This also explains why optical communication systems seem always to need microwatts of optical power, while your FM radio receiver operating at about 100 MHz gives a respectable signal-to-noise ratio for a few femtowatts (say, $1\,\mu V$ in a 75-Ω antenna). In each joule of radio photons there are five million times more photons than in a visible optical joule. We will return to this important point in Chap. 5, when dealing with detection noise.

The quantity $r(\lambda)$ is usually given in the photodetector manufacturers' literature. Figure 1.5 shows typical curves for some real photodiodes. The straight line is the ideal $0.807\,\lambda_{\mu m}$ result for a detector with 100 percent quantum efficiency, and it can be seen that the responsivity of real silicon diodes typically approaches within about 30 percent of the ideal from about $0.4\,\mu m$ to $1\,\mu m$. The ratio of actual responsivity to ideal responsivity is called the quantum efficiency (η):

$$\eta = \frac{r(\lambda)}{r_{ideal}(\lambda)} \tag{1.4}$$

Departures from the $0.807\lambda_{\mu m}$ unit quantum efficiency straight line for silicon diodes seen in Fig. 1.5 occur as a result of several effects. The rapid fall-off in sensitivity at wavelengths above approximately $\lambda_g \approx 1.1\,\mu m$ wavelength for silicon is caused by the increasing transparency of the silicon crystal at those wavelengths. Photons with energy less than the material bandgap energy E_g

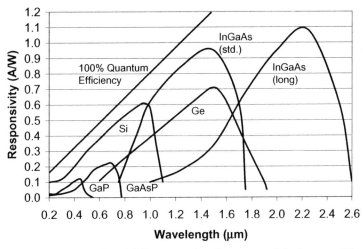

Figure 1.5　Photodiodes of different semiconductor materials show sensitivity in different wavelength regions, limited at long wavelength by their energy gap. 100 percent quantum efficiency means that one photon produces one hole-electron pair.

are simply not absorbed and hence pass through the crystal without being usefully detected. They serve only to warm up the back contact. The cutoff wavelength in microns is given approximately by $\lambda_g = hc/E_g \approx 1.24/E_g(V)$, where
$E_g(V)$ is the bandgap energy measured in electron-volts. Indeed, with a halogen
lamp, which emits both visible and near infrared light, and an infrared viewer
or camera sensitive to $1.3\,\mu$m you can see through a silicon wafer as if it were
glass.

Variations in doping can make minor changes in the bandgap and hence in
sensitivity at long wavelengths, either to increase it or to decrease it. In applications using the important neodymium laser sources emitting at around
$1.06\,\mu$m, even minor increases to cutoff wavelengths in silicon photodiodes can
be useful in increasing detection sensitivity. The peak sensitivity in conventional silicon diodes occurs at around $0.96\,\mu$m.

A reduction in long wavelength sensitivity is also sometimes useful, helping
to suppress detection of interfering infrared light, when low level signals at
visible and ultraviolet wavelengths are the target of interest. As we will discuss
in later chapters, the signals we want to see can often be swamped by signals
at other wavelengths. Tailoring the spectral sensitivity curve can bring great
advantages. For the high-volume applications of $0.88\,\mu$m and $0.95\,\mu$m LEDs,
used in handheld remote controls and IRDA short-range communications
systems, silicon photodiodes embedded in black filtering plastic are available.
The material is transparent in the near infrared but cuts out much of the visible
light below $0.8\,\mu$m, which greatly reduces disturbance from ambient light
sources.

The quantum photodetection process suggests that even short-wavelength
ultraviolet and x-ray photons should generate charge carriers in a silicon
photodiode. This is the case, but the component's detection process functions
only if the charge carriers are generated in or close to the depletion region. At
the short wavelength end of Fig. 1.5 the silicon is becoming too absorbing;
photons are being absorbed too close to the surface in a region where charge
carriers will not be swept away to contribute to the photocurrent. Again they
end up as heat. Improvements in ultraviolet (UV) sensitivity can be made
through careful control of detector doping, contact doping, and doping thicknesses to give a depletion region lying very close to the surface.

1.6   Other Detector Materials

Although important, silicon is not the only detector material that can be used
to fabricate photodetectors. The wavelengths you want to detect, and those you
would rather not detect, should give you an idea what energy gap is appropriate and hence what material is best. A few semiconductor energy bandgaps
are shown in Table 1.1.

As we have seen, to be detected the incident light must have a photon energy
greater than E_g (or a wavelength $\lambda < \lambda_g$). At wavelengths beyond $1.1\,\mu$m, where
silicon is almost transparent, germanium diodes are widely used and have been

TABLE 1.1 **Approximate Energy Bandgaps and Equivalent Wavelengths of Some Common Semiconductors**

Material	Bandgap energy (eV at 300 K)	Equivalent wavelength $\lambda_g(\mu m)$
C (diamond)	5.5	0.23
GaN	3.5	0.35
SiC	3.0	0.41
GaP	2.24	0.55
GaAs	1.43	0.87
InP	1.29	0.96
Si	1.1	1.11
$In_xGa_{1-x}As$	0.48–0.73	1.70–2.60
GaSb	0.67	1.85
Ge	0.66	1.88
PbS	0.41	3.02
PbTe	0.32	3.88

available for many years. They show reasonable responsivity out to almost 2 μm and have the big advantage of detection down to 0.6 μm. This allows experiments to be set up and their throughput optimized more conveniently with red light, before switching to infrared light beyond 1 μm. Although germanium covers the 1.3- to 1.6-μm region so important to fiber optic communications, this application is often better handled by another material, the ternary semiconductor indium gallium arsenide (InGaAs). Photodiodes formed of this material can have higher responsivity than germanium, and much lower electrical leakage currents. Recently a large choice of photodiodes formed in $In_xGa_{1-x}As$ has become available, driven by the fiber optic communications market. By varying the proportion x in the semiconductor alloy, the sensitive range of these devices can be tailored. In standard devices with $x = 0.53$ and bandgap $E_g = 0.73$ eV the response limits are 0.9 μm and 1.7 μm. By increasing x to 0.83 and changing $E_g = 0.48$ eV the response can be shifted to 1.2 to 2.6 μm. Figure 1.5 shows examples of both these responses. The advantage and disadvantage of InGaAs detectors for free space beam experiments are their lack of significant sensitivity in the visible, making visible source setups more difficult but cutting down interference from ambient light. Some help can come from the use of near infrared LED sources emitting at 0.94 μm, which are detected both by silicon and InGaAs devices.

Photodetectors are also available in several other materials. Gallium phosphide (GaP) offers a better match to the human eye response, especially the low illumination level scotopic response. We can even avoid the use of the correction filters which must be used with silicon detectors for photometric measurements. Gallium arsenide phosphide (GaAsP) is available both as diffused and as metal-semiconductor (Schottky) diodes and is insensitive above 0.8 μm. Hamamatsu has a range of both these materials (e.g., G1962, G1126). Opto Diode Corp. offers detectors of gallium aluminum arsenide (GaAlAs, e.g., ODD-45 W/95 W), which show a response strongly peaked at 0.88 μm, almost an exact match with commonly available 0.88-μm GaAlAs LEDs. This response greatly reduces the need for optical filtration with interference filters or IR-transparent black plastic molding for visible light suppression. This characteristic of being "blind" to interfering wavelengths is made good use of in detectors of silicon carbide (SiC). It is a high bandgap material that produces detectors with sensitivity in UV and deep blue ranges. They are useful in UV photometry and for blue flame detection. Silicon carbide photodetectors have been made available by Laser Components, which offer a peak sensitivity at 275 nm and a response that is very low above 400 nm. At the peak the responsivity is 0.13 A/W. Detectors fabricated from chemical vapor deposited diamond have also been described (Jackman 1996) for use from 180 to 220 nm. Sensitivity throughout the visible spectrum is insignificant. Last, a large range of lead sulphides, selenides, and tellurides are used for infrared detection in the 3- to 10-μm region.

For wavelengths below about 350 nm the normal borosilicate (Pyrex) glass used for detector windows becomes absorbing, and alternatives such as fused silica or synthetic sapphire must be considered. These are transparent to approximately 0.2 μm and 0.18 μm, respectively, depending on their purity and fabrication methods. Many manufacturers offer a choice of window materials for the same detector. At even shorter wavelengths, silicon can still be useful for detection, but the window must be dispensed with altogether. Some manufacturers provide windowless photodiodes in sealed, airtight envelopes. However, once the envelope is opened, maintaining the low electrical leakage properties of the photodiode under the attacks of atmospheric pollution and humidity is difficult. They gradually become much noisier. This should therefore be considered only as a last resort or where alternative protection can be provided. At these short wavelengths, air is itself becoming absorbing, necessitating vacuum evacuation of the optical path. This is the origin of the term *vacuum ultraviolet region.*

We have calculated the ideal responsivity of photodiodes assuming that the photon is absorbed in the depletion region. Another significant reduction in performance arises from photons that are reflected from the surface of the diode, never penetrating into the material, let alone reaching the depletion region. The fraction of energy lost in this way is given by the power reflection coefficients of the Fresnel equations (Fig. 1.6). For the simplest case of normal incidence of light from air into the material of refractive index n, the power reflectivity is $((n - 1)/(n + 1))^2$.

Detector semiconductors usually exhibit high refractive indices. For silicon with $n \approx 3.5$ the power reflectivity is 31 percent, leaving only 69 percent to penetrate into the detector material. To reduce this problem, detectors are often treated with antireflection (AR) coatings. For example, a one-quarter-wavelength ($\lambda/4$) layer of silicon nitride (Si_3N_4 with $n = 1.98$) can reduce the reflected power to less than 10 percent across the visible and near infrared and essentially to zero at a fixed design wavelength. For special uses, three or four photodiodes can be assembled to achieve very high absorption efficiency across

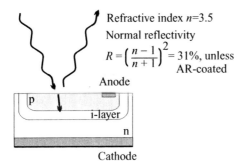

Refractive index $n=3.5$

Normal reflectivity

$$R = \left(\frac{n-1}{n+1}\right)^2 = 31\%, \text{ unless}$$
AR-coated

Figure 1.6 The high reflectivity of an air/semiconductor interface, given by the Fresnel equations, stops some incident light reaching the junction.

a wide wavelength band, with each detecting some of the remaining light reflected from the previous one. Alternatively, surface textures can be arranged to give a similarly high absorption. Where the light is incident from a transparent glass block or optical fiber, even index matching the fiber to the diode with a transparent gel or adhesive can roughly halve the reflection losses. To see this, substitute $n = 1.5$ for the "1" in Fig. 1.6.

1.7 Photodiode Equivalent Circuit

1.7.1 Current source model

To conveniently use the photodiode, we need a simple, didactic description of its behavior. The equivalent circuit we will use (Fig. 1.7) treats the photodiode as a perfect source of photocurrent in parallel with an ideal conventional junction diode. This is compatible with the physical model of Fig. 1.2. The photodetection process generates charge carriers and the internal photocurrent I_o. Note that we have no direct access to I_o. All we have is the external current I_p that is provided at the photodiode's output terminals. We showed earlier that under illumination the photodiode anode becomes positive. This tends to forward bias the pn-junction, causing internal current flow and a reduced output current.

Ignore for the moment the series resistance R_s, shunt resistance R_{sh}, and parasitic capacitance C_p. The output current is then given by I_o (calculated from the responsivity values discussed earlier) minus the diode current I_d flowing through the internal diode:

$$I_p = I_o - I_d \tag{1.5}$$

$$I_p = I_o - I_s(e^{qV_d/kT} - 1) \tag{1.6}$$

The second term is called the *Shockley equation*, the expression relating current and voltage in an ideal junction diode. The new parameters are as follows:

k: Boltzmann's constant $(1.381 \times 10^{-23}\,\mathrm{W \cdot s/K})$

T: Absolute temperature (about $300\,\mathrm{K}$ at room temperature)

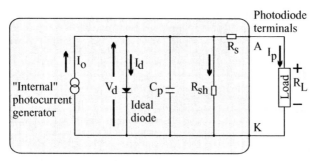

Figure 1.7 A photodiode can be modeled as a current generator proportional to the incident light intensity in parallel with an ideal diode, a shunt resistance, and a shunt capacitance. These significantly influence the diode's performance, depending on the external circuitry.

V_d: Junction voltage (V)

I_s: Reverse saturation current (A), a current that we hope is much smaller than the photocurrents we are interested in detecting.

Depending on how the diode is connected to the external circuit, V_d and I_p can take on widely different values for the same illumination power. Two simple cases are easy to solve and useful in practice.

1.7.2 Open circuit operation

If the diode is operated open circuit, or is at least so lightly loaded that the external current I_p is negligible, then all the photocurrent flows through the internal diode. The above equation then gives:

$$I_o = I_s(e^{qV_d/kT} - 1) \tag{1.7}$$

or:

$$V_d = \frac{kT}{q} \ln\left(1 + \frac{I_o}{I_s}\right) \tag{1.8}$$

The junction voltage, and hence also the open circuit terminal voltage, is therefore an approximately logarithmic function of the incident power. Putting in real values for k, q, and $T = 300\,\mathrm{K}$ we can calculate the term $kT/q = 0.026\,\mathrm{V}$. Hence if $I_o/I_s \gg 1$, then each decade increase in internal current increases junction voltage by $\ln(10) \cdot 26\,\mathrm{mV} = e \cdot 26\,\mathrm{mV} = 60\,\mathrm{mV}$. The logarithmic response is useful for measuring signals with widely varying intensities and for approximately matching the response of the human eye. However, accuracy is usually not high.

1.7.3 Short circuit operation

If, alternatively, the diode is operated into a short circuit, we have $V_d = 0$. With no voltage, no current can flow through the internal diode, and the full internally generated photocurrent is available at the output terminals. Then the bracketed quantity in Eq. 1.6 is zero, and internal and external currents are equal. Now the external current is linearly related to the incident power. This linearity can hold over at least 6 or even 10 orders of magnitude of incident power. At the very lowest detectable photocurrents, that is, for a small ratio of I_o/I_s, the presence of I_s cannot be neglected. At the other end of the scale photodiodes cannot handle arbitrarily large photocurrents. At high currents the photodiode series resistance shown in Fig. 1.7 contributes a voltage drop, diverting some of the internal photocurrent through the diode and through R_{sh}. Hence linearity can suffer. Fine wire bonds can even be melted. Photodiodes are usually specified with limits either on total power (mW) or power density (mW/mm^2). With focussed laser sources these limits can easily be reached, leading to odd behavior.

Equations (1.7) and (1.8) suggest that the photocurrent decreases and the open circuit voltage increases with increasing temperature. This is not the case experimentally observed. For example, the data sheet of the popular BPX65 photodiode (manufactured by Infineon, part of the Siemens group, and others) shows a short circuit temperature coefficient for external current of the order of +0.2 percent/°C. This can be explained by noting that reverse saturation current is also an exponential function, increasing with temperature:

$$I_s \approx e^{-E_g/kT} \tag{1.9}$$

1.7.4 General operation

Apart from the open and short circuit configurations, the photodiode may in general be operated with a finite load resistor varying over a wide range of values and with an applied voltage in forward or reverse bias. Hence it is useful to be able to calculate the output current and voltage under any such conditions. This is a little trickier than for the open and short circuit conditions; one approach is given here. In Fig. 1.8 we generalize the circuit and add a bias voltage source V_b in series with the load resistor R_L. Then we can write two expressions for I_p, one from the internal diode components and one from the external circuit:

$$I_p = \frac{V_d - V_b}{R_L} \tag{1.10}$$

$$I_p = I_o - I_d \tag{1.11}$$

$$= I_o - I_s \cdot (e^{q \cdot V_d/k \cdot T} - 1) \tag{1.12}$$

Changing I_s to I_s' to give a more realistic temperature dependence of Eq. (1.9):

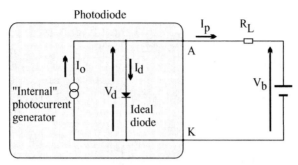

Figure 1.8   The photodiode's terminal voltage can be obtained by solving the circuit equations given the external bias voltage and load resistor.

$$I_p = I_o - I_s'e^{-qE_g/kT}(e^{qV_d/kT} - 1) \tag{1.13}$$

Equate the two expressions and rearrange:

$$F(V_d) = I_o - I_s'e^{-qE_g/kT}(e^{qV_d/kT} - 1) - \frac{V_d}{R_L} + \frac{V_b}{R_L} = 0 \tag{1.14}$$

To solve this for V_d we can either plot the function and just look for the zero or use Newton's method to iteratively generate new estimates of V_d:

$$V_{d_{z+1}} = V_{d_z} - \frac{F(V_{d_z})}{F'(V_{d_z})} \tag{1.15}$$

where $F'(V_d)$ is the derivative of $F(V_d)$ with respect to V_d. With a few iterations, this usually converges on a solution for the diode voltage V_d and hence for the other currents and voltages. With the mathematical software Mathcad®* we can just solve for V_d using something like $V_d = 0$, solution:= root($F(V_d),V_d$). Note that this works for forward and reverse bias, with and without photocurrents, and so can be used in all three quadrants of the photodiode characteristics.

Figure 1.9 shows the complete schematic current/voltage characteristic of a photodiode under three levels of illumination. In the first quadrant (positive voltage and current) the curve labeled "dark" is just the exponential forward characteristic of a junction diode. In the third quadrant only a very small current flows (I_s). As the level of incident light is increased, the curves shift bodily downward in the negative current direction. This shift is linear in incident power, as is the set of points on the $V = 0$ axis marked I_{SC} (short circuit). The zero current points marked V_{OC} (open circuit) are clearly a nonlinear function of illumination, as discussed earlier. At very high reverse bias voltages the current may increase rapidly to large and possibly damaging values. This is the reverse breakdown region.

---

*Mathcad is a registered trademark of Mathsoft Engineering and Education Inc.

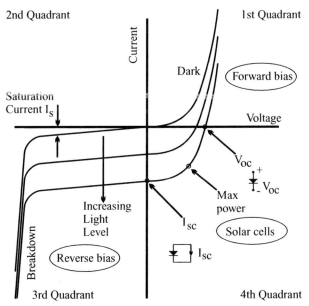

Figure 1.9 Four-quadrant current/voltage characteristic. Without illumination this is similar to a conventional diode. Increasing illumination shifts the characteristic in the negative current direction. Detection is possible in quadrants 1, 3, and 4.

In the fourth quadrant we can extract power from the detector. This is the region where solar cells are designed to work. For best efficiency we should define the operating point through choice of the load resistance to maximize the VI product. At this point it is possible to extract roughly 80 percent of the $I_{SC}V_{OC}$ product. Solar cells are pn-junction devices, usually designed with higher doping levels than pin-photodetectors to keep series resistance low and to maximize the absorption of light at the short wavelength end of the silicon absorbance spectrum. This is to best match the spectrum of sunlight. Although solar cells are rarely considered as detectors for instrumentation, their low cost/area can make them very attractive as such.

1.7.5 Parasitic capacitance

Let's look now at the other elements of our equivalent circuit of Fig. 1.7. The parasitic capacitance which appears as if across the photodiode is C_p. Its origin lies in the positive and negative space charges separated by the depletion region of Fig. 1.2, which act like a parallel plate capacitor. Although the diode area is rather small compared with a component capacitor, the thinness of the depletion region can lead to high capacitance values. It increases approximately linearly with the detector area and decreases with increasing reverse bias. Measurement of the junction capacitance as a function of reverse voltage can tell us the device's doping profile. A plot of $1/C^2$ versus voltage can provide the junction bandgap energy (Sze 1969). Different processing parameters during manufacture can have a large effect on C_p. In Chaps. 2 and 3 we will see how important is C_p to the speed and noise performance of an amplified photodetector.

1.7.6 Leakage resistance

All photodiodes act as though they have a resistance shunting the diode (R_{sh} in Fig. 1.7). Its value depends on the processes used to fabricate the diode substrate and junction and decreases with increasing area. A high quality 1-mm^2 silicon device (e.g., PIN040 from UDT Sensors Inc.) specifies a leakage resistance of typically $1\,G\Omega$ and not less than $200\,M\Omega$ at room temperature. Temperature increase reduces R_{sh}. Other manufacturers publish instead a leakage current or dark current, which is just the current that will flow through the shunt resistance for a specified terminal voltage. If the diode voltage in the actual circuit is zero, then no current will flow and the effects of R_{sh} will be minimized. We will see later that R_{sh} can have a significant effect on the detection signal-to-noise ratio. For low values of R_{sh} even a few millivolts offset from an operational amplifier can lead to significant leakage current.

Leakage can be greatly modified by process parameters, although significant reduction is usually at the expense of other diode characteristics. For example, the 7.6-mm^2 detection area BPW34 photodiode manufactured by Siemens is specified with a leakage of $2\,nA$ at 10-V reverse bias and capacitance of $72\,pF$ at zero bias. The very similar BPW33 exhibits only 20-pA dark current, but the parasitic capacitance increases to $750\,pF$ under the same conditions.

1.7.7 Series resistance

Series resistance comes from the bulk resistance of the photodiode substrate, from the ohmic contact diffusions, and from the leads. It is probably the least troubling diode characteristic for normal laboratory detection needs but becomes important in high bandwidth systems and in power generators using solar cells.

1.7.8 Table of representative photodiodes

Table 1.2 shows a few representative photodiodes. The BPW34 is a common silicon device in a clear plastic package and has been available for 30 years. The BPW33 looks similar, with the same active area. However, as noted, the dark leakage current has been reduced significantly, at the expense of a tenfold increase in junction capacitance. Items 3 and 4 are modern design silicon devices and show the change in dark current and capacitance incurred by changing the active area. These also offer very low temperature coefficients of responsivity at less than $100\,ppm/°C$. Items 5 and 6 are InGaAs detectors. The first has an area of $1\,mm^2$ and a capacitance of $90\,pF$. Item 6 is similar, except that a spherical lens has been used to accept about $1\,mm^2$ of beam area but focus it onto a

TABLE 1.2   Selection of Representative Photodiodes

	Type	Active Dimensions	Peak Respons. (A/W)	Dark Current	Capacitance (pF)	NEP (W/Hz$^{1/2}$)
1	Infineon BPW34	2.75 × 2.75 mm	0.6 @ 0.85 μm Silicon	2 (<30 nA) @ 10 V	72 pF @ 0 V	4.2 × 10^{-14}
2	Infineon BPW33	2.75 × 2.75 mm	0.6 @ 0.85 μm Silicon	20 pA @ 1 V	750 pF @ 0 V	5.3 × 10^{-15}
3	Hamamatsu S1336-18	1.1 × 1.1 mm	0.5 @ 0.95 μm Silicon	20 pA @ 10 mV	20 pF @ 0 V	5.7 × 10^{-15}
4	Hamamatsu S1337-1010	10 × 10 mm	0.5 @ 0.95 μm Silicon	200 pA @ 10 mV	1100 pF @ 0 V	1.8 × 10^{-14}
5	Hamamatsu G8370-01	1-mm diam.	0.95 @ 1.55 μm InGaAs	5 nA @ 5 V	90 pF @ 5 V	2 × 10^{-14}
6	Hamamatsu G6854-01	(Spherical lens)	0.95 @ 1.55 μm InGaAs	80 pA @ 5 V	1 pF @ 5 V	2 × 10^{-15}
7	AME-UDT BPX65	1 × 1 mm	0.55 @ 0.90 μm Silicon	500 pA @ 20 V	3 pF @ 20 V	2.3 × 10^{-14}
8	AME-UDT InGaAs-100L	100-μm diam.	0.95 @ 1.55 μm InGaAs	50 pA @ 10 V	1.5 pF	—
9	Hamamatsu G8198-01	40-μm diam.	0.95 @ 1.55 μm InGaAs	60 pA @ 5 V	0.6 pF @ 2 V	2 × 10^{-14}
10	OSI Fibercomm FCI-InGaAs-25C	25-μm diam.	0.95 @ 1.55 μm InGaAs	100 pA @ 5 V	0.2 pF @ 5 V	—

much smaller diode. In the process the capacitance has dropped to 1 pF. Item 9 is another device manufactured for a very small surface area and for use with single-mode fibers. Subpicofarad capacitance has been achieved in this way. Item 10 is a small active area and low capacitance chip device offered for 10-Gbps telecomms applications. The last column gives the noise equivalent power (NEP) at the wavelength of peak responsivity. This is the lowest power that could be detected under optimum conditions in a 1-Hz bandwidth, limited by the dark current noise.

1.8   APDs, Photomultipliers, Photon-Counting

Avalanche photodiodes (APDs) are similar to conventional pin photodiodes but are designed to produce and withstand a high internal electric field, which can accelerate photoelectrons generated near the depletion region. As they are accelerated, additional photoelectrons are generated and a cascade process produces 10 to 1000 photoelectrons per incident photon. This high internal gain can be of great use in receiver design. The gain is produced through application of a higher than normal reverse terminal voltage. For example, the low-bias Hamamatsu S2383 is a 1-mm-diameter silicon device that operates at 150 to 200 V for a gain of 100. Some very large APDs also are now becoming available. Advanced Photonix Inc. offers, for example, the 16-mm diameter 630-70-72-

5X1, providing a responsivity of 93 A/W at 750 nm with a bias voltage in the range of 1700 to 2000 V. APDs in InGaAs are also available for long wavelength (1.55 μm) use.

Note that these detectors are not inherently *more sensitive* nor do they offer *lower noise* than conventional pin-diodes. They all offer about the same wavelength response, capacitance, and quantum efficiency and so have essentially the same limitations. APDs are also inherently noisier because of the presence of excess noise caused by fluctuations in the gain mechanism. However, where the provision of sufficient external electronic gain and bandwidth is problematic, leading to limitations in receiver performance, the internal gain of the APD can be very useful. This is generally the case only at higher speeds (say 1 MHz and above). They are rarely superior for low light level detection at low frequencies. They are also much more difficult to use than pin diodes, owing to the precise regulation of the high bias voltage that is needed, which is itself temperature-dependent. However, complete detection modules are available from Hamamatsu and others to ease the design difficulties.

Similar tradeoffs occur with photomultipliers (PM tubes), which generate photoelectrons in an inorganic semiconductor phosphor. Like the APDs, these use an internal cascade multiplication process to achieve very high external responsivities, typically 10^5 to 10^7 A/W. However, their intrinsic sensitivity is rather low, as their photocathodes exhibit quantum efficiencies much less than peak semiconductor detector values, for example, 0.1 A/W. They also require a stable, low noise high voltage supply at the kilovolt level. Advantages of PM tubes are their wide range of sensitivity obtained through a choice of photocathode materials, large detection areas compared with semiconductor photodiodes, and low dark currents. They are also relatively insensitive to ionizing radiation. Both APDs and PM tubes are available with low enough dark currents to be used in photon-counting mode. Photon counting is a sensitive detection method for use at low incident powers (typically $<10^9$ photoelectrons/second), where instead of averaging photoelectrons into a continuous current, individual electron charges are counted. This brings the advantage of a kind of separation between thermal noise and photoevents through the use of carefully set up threshold discriminators. Large gains in performance are sometimes possible using this mode of detection.

1.9 Summary

In this chapter we have discussed the basics of photodetection using junction photodiodes and shown how they can be represented by a simple current-generator model. This is adequate in most cases to predict performance in a wide variety of externally connected circuits. Different materials may offer advantages in special situations, but the majority of laboratory and instrumentation applications are dealt with using silicon and InGaAs devices. With this information, we will now investigate how to use photodiodes in practice to perform useful measurements.

2

Amplified Detection Circuitry

2.1 Introduction

In this chapter we progress from photodetectors to photodetection, simple electronic circuits that allow observation and measurement of static and varying optical signals with a voltmeter, on an oscilloscope, or as part of an optoelectronic product. We saw in Chap. 1 that the signal output from a photodiode is strongly influenced by the circuitry connected to it. In particular, unless the load voltage is much smaller than a few tens of millivolts, simple connection to a resistive load leads to a logarithmic or at least highly nonlinear response, while reverse biasing into the third quadrant of the diode's IV characteristic generally leads to a current-output response that is linear over many orders of magnitude of incident power. This reverse bias connection is the basis of much optical measurement technology. In fact, the simple circuit seen in Fig. 1.8 can be called our first "instrument."

2.2 The Bias Box

Figure 2.1 shows a photodiode reverse biased by a small battery, with a series load resistor whose imposed voltage can be read with a high impedance voltmeter or oscilloscope. With the addition of a multiway switch to change load resistors over a wide range, say from $100\,\Omega$ to $1\,M\Omega$, and perhaps a changeover switch to swap the detector polarity, this "equipment" should be in every optical researcher's kit bag. I try to keep several available. A 3.3-V lithium cell or 6-V camera battery is unlikely to damage the majority of silicon photodiodes through reverse breakdown, but check the photodiode's voltage ratings first to be sure. However, it is possible to pass excessive current if the photodiode is brightly illuminated on the low load resistance settings, so connect it first, monitor the voltage, and then illuminate. The bias box can be used with almost any detector that is handy. It has an easily varied sensitivity and responds linearly to power changes. It never oscillates, and the battery voltage is almost

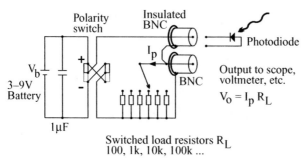

Figure 2.1   The bias box uses a small battery to reverse bias a photodiode and pass its photocurrent through a choice of load resistors. It is a simple, useful detector. Changing the resistor allows a trade-off between output voltage and detection speed.

always sufficient. The battery should last for months if the detector is not left in the sunlight.

However, this circuit has limitations. In using it with an oscilloscope you will find that the most sensitive range ($1\,\mathrm{M\Omega}$) gives less output voltage than expected, because of shunting of the bias box load with a second 1-$\mathrm{M\Omega}$ resistor in the scope. You may also have problems with voltage offsets, even in total darkness, caused by leakage currents driven by the reverse voltage. There may also be pickup of electrical interference around the photodiode's floating BNC socket. You *can* connect the photodiode with a length of coaxial cable, but this is not recommended. The cable is floating too, so touching the cable screen against the output BNC screen under incorrect forward bias may destroy the diode through excess current. Pickup also will be worse, and the cable capacitance appears directly across the diode, limiting high-speed response. The bias box is extremely useful for the initial investigations, but once the approximate optical signal power level is known and the performance requirements are better defined, using an optimized amplified detection circuit is preferable. Several combinations of photodiodes with electronic amplifiers are useful for your arsenal.

2.3   Voltage Follower

For many applications with low light levels a 1-$\mathrm{M\Omega}$ load resistor is too small to optimize the signal to noise. To overcome the input impedance limitation of the scope, a simple impedance buffer or *voltage follower* may be used. This can easily be configured with an operational amplifier (opamp) as shown in Fig. 2.2. Connected like this the opamp has a voltage gain of unity and a high input impedance limited by its input bias current. If a field effect transistor (FET) amplifier is used, the effective impedance can be several gigohms at DC. For a linear dependence of the output voltage on light intensity, reverse bias of the photodiode will still be needed, just as in the bias box. Bias voltage must be

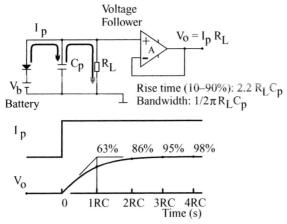

Figure 2.2 The voltage rise time from a biased photodiode is proportional to the total shunt capacitance, which includes the diode's junction capacitance as well as any circuit capacitance.

large enough to maintain a reverse bias, let's say at least a volt, even if reduced by the passage of photocurrent through R_L. This circuit is therefore a bias box with voltage buffer to isolate the high photodiode load from a low-impedance oscilloscope or voltmeter input. The output voltage at DC is as before just the photocurrent I_p flowing through the load resistor R_L:

$$V_o = I_p R_L. \tag{2.1}$$

This circuit allows use of load resistors much greater than $1\,\mathrm{M\Omega}$. The limit comes, as with the bias box, when the voltage dropped across the load resistor becomes comparable with the bias voltage (V_b), whatever its origin. The load resistor voltage is due not only to photocurrent, but also to photodiode "dark" leakage currents and to the input bias current of the amplifier.

2.3.1 High-value load resistors

In some applications it is desirable to use very large values of load resistor, for example, $100\,\mathrm{M\Omega}$ to $100\,\mathrm{G\Omega}$. In these cases only amplifiers with the lowest bias currents can be used. Amplifiers with bipolar transistor front ends typically need bias currents of the order of $1\,\mu\mathrm{A}$, and so are excluded from these applications by the large offset voltages they would produce. Amplifiers with junction FET input stages, such as the popular LF356 series, require bias currents of the order of $200\,\mathrm{pA}$, and so can perform better. Amplifiers with metal oxide semiconductor FET (MOSFET) input devices, such as the CA3140, offer input bias currents of the order of a few tens of picoamperes down to a picoampere, and so can be used with gigohm resistors without excessive offsets. Specially optimized amplifiers, such as the Analog Devices AD515 ($\pm 75\,\mathrm{fA}$), Burr-Brown

OPA128 (±75 fA), and National Semiconductor LMC6001A (±25 fA) reduce bias errors even further. These would allow the use of $100\,G\Omega$ resistors with only 2.5-mV DC offset. Note, however, that these FET bias currents exhibit a rapid increase with temperature, which for high temperature operation may even exceed the current of some bipolar devices. This must be considered in the full design of a receiver subjected to high temperatures.

2.3.2 Guarding

At these high impedance levels, other currents can play a big role. Leakage current flowing along the surface of printed circuit boards (PCBs) is often significant and is greatly increased by humidity films and surface contamination. It can be reduced through the use of guard rings. These are electrodes that surround the sensitive input pins and are either grounded or driven at low impedance at the same voltage as the input pins. If the voltages of pin and interfering electrode are identical, by definition no current can flow. A voltage-follower can be used to drive the guard electrode to the input voltage, to an accuracy limited by the amplifier voltage offset errors. To understand how guard rings work, it is really necessary to "think in 3-D" (Fig. 2.3a). The pin shown passing through the PCB is the sensitive point, for example connecting a low-current photodiode to the front end of a sensitive receiver. There is also an interfering electrode close by, at a source potential V_{source}. Imagine that the PCB is coated in a 1-mm-thick layer of weakly conducting material, an electrolyte that couples the interfering electrode and sensitive pin. To work out where the currents flow, we really need to perform a 3-D finite element analysis, when it would be seen that volume currents flow in the full thickness of the conducting layer, part of that current flowing to the sensitive input pin. If now we add a guard electrode which surrounds the sensitive pin, preferably on both sides of the PCB, *some* of the volume currents will flow to the guard electrode, depending on its potential, size and geometry in relation to the other electrodes. If the area of the guard electrode is too small, stray currents will bypass the guard and make it to the input pin. The greater the area of the guard electrode, the more stray current will be mopped up. Conversely, the thicker the conducting layer, the more stray leakage current will pass by the guard electrode to the sensitive pin. Hence we cannot say how big the guard electrode should be without having an idea of the thickness of the conducting layer. Common sense suggests that the guard ring width should be at least as great as the conducting film thickness.

As shown, current leakage through the bulk of the PCB material can also be a problem, which is hardly reduced by *surface-printed* guard rings. Special copperclad boards made from PTFE polymers or ceramics are available which reduce this leakage contribution, but these are rarely seen in one-off designs. For test lash-ups it is not hard to make custom boards using sheet PTFE, subsequently carefully cleaned, PTFE integrated circuit sockets, PTFE insulated terminals or just air! In principle, *volume* guard rings could also be fabricated,

(a)

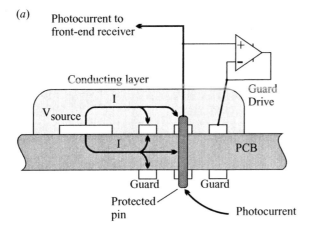

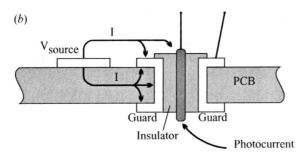

Figure 2.3 Guarding using printed electrodes can only protect against surface currents (a). Currents may also flow in the bulk of the printed circuit board and in any conducting layer. The calculation of the currents requires a 3D model. Fully coaxial guarding (b) can additionally reduce PCB currents.

with an isolated coaxial-structure electrode similarly driven at the same potential as the sensitive input pin. The structure depicted in Fig. 2.3b could be fabricated using large-diameter plated through-holes with pressed-in PTFE inserts. It is still necessary to minimize the conductance of the insulating region, for example by using a PTFE insert, and to carefully choose the size of the guard electrode. With the low conductance of PTFE and only a few millivolts or less of voltage stress, stray currents in this configuration are expected to be very low. Note that even with air insulation, there will still be leakage currents from the occasional cosmic ray strike, which are energetic enough to ionize oxygen and nitrogen ($\approx 12\,\mathrm{eV}$) and provide charge carriers. This suggests using the smallest volume of insulation possible. Clearly a compromise must be reached. The level of effort of volume-guarded PTFE inserts is probably justifiable only in specialized test equipment, for example to routinely measure with low fractional errors the bias currents of the highest performance amplifiers. A fascinating discussion of such experiments has been published by Pease (1993).

For the highest impedance designs, it is often better to dispense with integrated circuit (IC) sockets and circuit boards altogether. ICs can be directly soldered to cleaned PTFE stand-off terminals that are hand wired, or a compromize is to use a conventional IC socket with the critical opamp input pins lifted clear and soldered directly to the photodetector and load resistor. A disadvantage of this construction technique, apart from its time and expense, is its performance under acceleration. Vibration will vary the mechanical separation of components, modulating parasitic capacitances and causing microphonic pickup.

2.3.3 Cleanliness and flux residues

Some types of solder fluxes seem to be almost conducting! I have done no quantitative tests, but the water-soluble fluxes contained in "green" water-washable solders seem to be the worst offenders. Hence all flux and contamination must be carefully washed away from PCBs before use. In the past fluorocarbon solvent vapor cleaning systems were used for this. More environmentally friendly aqueous cleaners should now be used. These are available for large tank use or as small aerosols. They must in turn be followed by thorough rinsing in deionized or low-conductivity water, with the outlet water checked using a good conductivity meter to be sure that it approaches the limiting water value ($\approx 18 \, M\Omega \cdot cm$).

For hobby or occasional use diligent scrubbing of solder residues with an old toothbrush and methanol, isopropanol, a specifically designed aqueous cleaner, or even household detergent, followed by long rinsing under running tap water, can be almost as effective. In every case the board should be shaken or blown clean of excess water and allowed to dry thoroughly in warm air before use. Hair dryers and airing cupboards are key equipment here! Failure to remove these residues can lead to very strange behavior of detector circuits.

2.4 Big Problem 1: Bandwidth Limitation Due to Photodiode Capacitance

A significant problem with high-value load resistors is their very sluggish response to incident optical power changes. As Fig. 2.2 shows, the photodiode parasitic capacitance (junction + packaging + wire connections), along with the amplifier input capacitance, appear in parallel with the load resistor. Hence for the voltage on C_p (and therefore on R_L) to change, the capacitance must be charged and discharged. Because the photodiode operates as a current generator, this voltage change takes time.

For a step change in input power, the output voltage exponentially approaches its final value, reaching in one $R_L C_p$ time constant 63 percent $(1 - e^{-1})$ of final value. If the photodiode is a hardly biased BPW33 with a capacitance of 720 pF, and $R_L = 1 \, G\Omega$, the time constant is 0.7 seconds. Hence 95 percent $(1 - e^{-3})$ of final value is achieved only after more than 2 seconds. Stated alternatively, the photoreceiver bandwidth is approximately $1/2\pi R_L C_p = 0.22\,\text{Hz}$. This slow performance may be unacceptable for the application.

2.4.1 Reverse bias

The simplest change that can often be made to improve the bandwidth of an existing photoreceiver design is the application of photodiode reverse bias. All photodiodes may be reverse biased to some extent, and in some cases the gains can be very worthwhile. For example, the well-known BPX65 is a 1-mm^2 silicon device designed for high-speed applications. With zero bias its capacitance is about 15 pF. However, this device can withstand 50-V reverse bias (Fig. 2.4), at which point the capacitance will drop to about 3 pF. This can make a critical improvement in detection bandwidth. Most telecommunications photodiodes are designed to be operated in this mode, some being capable of withstanding 100 V or more.

Unfortunately, a bigger reverse bias will increase the reverse leakage current. In every case it is important to calculate and measure both the DC offset and the extra shot noise caused by the dark current. It is often convenient to make the reverse bias potentiometer-adjustable, so that the optimum bandwidth and noise setting can easily be found. In some cases even 1 V reverse bias can make a useful improvement in overall performance, without unduly increasing noise.

2.4.2 Photodiode choice

When reverse bias does not achieve the desired speed performance, try alternative devices. A very wide variety of photodiodes is available, which vary in size from 1000 mm^2 tiles of silicon to tiny 25-μm diameter devices designed for

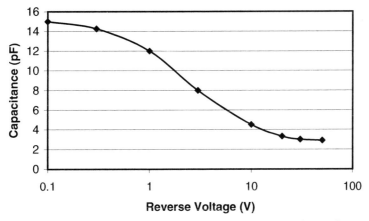

Figure 2.4 Photodiode capacitance is a function of reverse bias voltage. Values for a BPX65 are shown. High voltage will increase detection bandwidth at the expense of increased dark leakage current.

coupling to optical fibers. (see Table 1.2 for a few examples.) When speed is the key parameter, and all the light can be collected by the diode, the smallest possible device should be selected. For example, in a typical single-mode fiber communications receiver, infrared light is guided in a fiber core with a mode diameter of the order of $10\,\mu m$. If the coupling tolerances can be handled and the fiber can be positioned close enough to the chip, this is as large as the photodiode needs to be (Fig. 2.5a). Photodiodes are available with a diameter of $25\,\mu m$; this is small, but the area is still approximately 10 times bigger than necessary. Although at these small sizes the capacitance is unlikely to reduce proportionately to area, some gain in improvement is inevitable with smaller devices.

2.4.3   Optical transforms

We could in principle do even better with optical matching of source and detector. Liouville's theorem states that the étendue of an incoherent optical system cannot be reduced, but it can at least be manipulated. The étendue can here be thought of as the product of the beam's area and the square of the numerical aperture (NA^2). The light's NA on exiting the fiber is only about 0.12 and the mode area is about $100\,\mu m^2$, giving a product of $1.4\,\mu m^2$. Now, unless restricted by packaging, the photodiode can accept light over its full area and almost 2π steradians, about $0.5NA^2$. Hence if we can fill this acceptance NA, it can be traded for a reduction in spot size. We can use a microoptic lens system to do this (Fig. 2.5b), forming a spot of approximately $1.6\,\mu m$ diameter. This is as big as the photodetector needs to be. The capacitance of such a small detector could be far less than that of conventional, even very small devices. If this could be used without being swamped by transistor and packaging capacitance, power lost in high-angle Fresnel reflection coefficients or in positioning tolerances, performance improvements would be expected. It would be most convenient if the microoptic lens were integrated permanently with the tiny photodiode, for illumination from the single-mode fiber or fused onto the fiber end. At least one company (ALPS Electric) offers lenses and lensed detectors with this

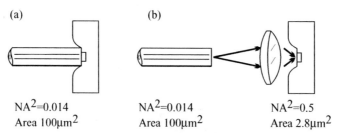

(a)   (b)

$NA^2=0.014$ $NA^2=0.014$ $NA^2=0.5$
Area $100\mu m^2$ Area $100\mu m^2$ Area $2.8\mu m^2$

Figure 2.5   Careful alignment of single-mode optical fibers allows the use of photodiodes with very small areas ($<30\,\mu m$ diameter). Increasing the numerical aperture (NA) of the incident light in principle allows the use of even smaller detectors.

improvement in mind. Similar gains could be obtained using NA-transforming fiber tapers.

Another application where lens systems are important is a free-space system such as burglar alarm systems, industrial control beam-sensors, and terrestrial and intersatellite free-space communication systems. In these cases there is little divergence in the incident light as it arrives, and it is spread beyond the area of any conventional photodetector. Hence a collection lens can give large signal increases without additional noise. The only drawback is that the reduced acceptance angle of the lensed receiver requires more accurate pointing and mechanical stability or active direction control. For intersatellite communications, much system complexity comes from the scanned acquisition of the transmitted signal at a relatively wide receiver acceptance angle, followed by active focus control to optimize received power. Where the apparent light source is large and indeterminate in arrival direction, for example in a domestic diffuse light communication system, it may be more efficient to use large detectors (solar cells have the lowest cost per unit area) or detectors made to "look" larger by embedding in a transparent hemisphere of high refractive index.

2.5   Transimpedance Amplifier

2.5.1   Why so good?

When the best, lowest capacitance devices have been chosen and reverse bias still does not provide the bandwidth required, another way to speed enhancement is to use the transimpedance amplifier configuration (Fig. 2.6). The transimpedance amplifier uses the same opamp as in Fig. 2.2, with the same load

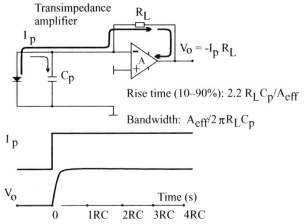

Rise time (10–90%): $2.2\, R_L C_p / A_{eff}$

Bandwidth: $A_{eff}/2\pi R_L C_p$

Figure 2.6 The transimpedance configuration gives the same output voltage as a biased load resistor, but a parasitic capacitance-limited rise time reduced by the effective gain of the amplifier.

resistor, but arranged differently. The opamp is connected with resistive feedback provided by the load resistor and the noninverting input is grounded. Negative feedback tries to force the two amplifier inputs to the same potential, so the inverting input becomes a "virtual earth." The impedance seen looking into this node is low.

Because the amplifier's input current is very low, a few tens of picoamperes, the bulk of the photocurrent has to flow as shown through R_L to the amplifier's output. With the polarities shown the output voltage must therefore become negative to "pull" current out of the photodiode anode. With the capacitance connected to the virtual earth, changes in photocurrent barely change the voltage on C_p. If the voltage does not change then neither does the charge $Q = CV$, and its apparent capacitance is greatly reduced. In the ideal case the diode capacitance is effectively shorted out, making it invisible to the photocurrent and feedback resistor. Consequently, the slow response time of the follower is significantly improved.

It is as though the capacitance now experiences a feedback resistance reduced by the value of the effective loop gain of the amplifier. The limiting bandwidth of the receiver (f_{limit}) is then given by the original bandwidth of the bias box or follower circuit ($1/2\pi R_L C_p$) multiplied by the open loop amplifier gain *at the limiting bandwidth*. Hence we can write:

$$f_{\text{limit}} = \frac{A_{\text{flimit}}}{(2\pi R_L C_p)} \tag{2.2}$$

where A_{flimit} is the closed loop amplifier gain at the limiting frequency. The majority of conventional, frequency-compensated opamps use an internal RC combination to give a dominant frequency pole at around $f_1 = 20\,\text{Hz}$ (Fig. 2.7). Above this frequency the gain drops off at a rate of −20 dB/decade, reaching 0 dB (unity gain) at the frequency corresponding to the "gain-bandwidth product" (GBW) or "unity gain frequency." The gain is therefore an approximately inverse function of frequency over much of the useful frequency range and GBW/f_1 = DC gain. In the figure, $20 \cdot \log(4\,\text{MHz}/20\,\text{Hz}) = 106\,\text{dB}$, and the gain at any frequency f is approximately GBW/f. Hence we can write:

$$f_{\text{limit}} = \frac{\text{GBW}}{(f_{\text{limit}}\, 2\pi R_L C_p)} \tag{2.3}$$

Rearranging the above for the upper frequency limit or bandwidth of the photodiode-transimpedance amplifier configuration we obtain:

$$f_{\text{limit}} = \left(\frac{\text{GBW}}{2\pi R_L C_p} \right)^{1/2} \tag{2.4}$$

This approximate expression should not be relied on for exacting accuracy, but it is useful for this situation. Typically the limiting frequency will be about one-half the value calculated from Eq. (2.4). For example, the LMC7101 FET opamp

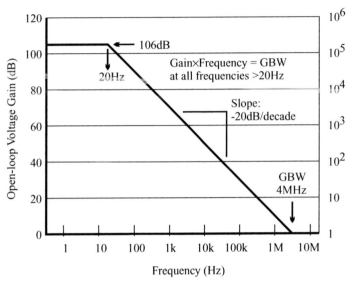

Figure 2.7 The gain of a conventionally compensated opamp is an inverse function of frequency. Gain bandwidth (GBW) is the frequency at which the gain is unity (0 dB).

is specified with GBW = 0.6 MHz. Evaluating the transimpedance configuration with BPW33 photodiode and 1-GΩ load, we calculate a bandwidth of 363 Hz, which is significantly better than the $1/(2\pi\, R_L C_p) = 0.22$ Hz of the follower configuration.

2.5.2 Instability

Certain practical details must be considered in the use of the transimpedance configuration. People are sometimes heard to say "transimpedance amps never work properly. They always oscillate." This is not entirely false! An opamp with resistive feedback and significant capacitance at the inverting input *should oscillate*, usually at a frequency around its unity gain frequency. This is because the extra phase shift caused by the low-pass $R_L C_p$ feedback is added to the amplifier's own phase shift. At some high frequency it will probably lead to positive feedback; if the gain is above unity it will oscillate. One solution is to add a small capacitance in parallel with R_L, to reduce the transimpedance at high frequencies.

For transimpedance amplifiers with $R_L > 1\,\text{M}\Omega$ resistance and small photodetectors, the value of capacitance needed is typically at the very low end of available capacitor ranges, a few picofarads or less. One effective approach to frequency compensation is to use a home made capacitance by tightly twisting two 30-mm lengths of fine enameled or plastic-insulated solid wire (30AWG, or any wire-wrap wire) and soldering the ends to the feedback resistor (Fig. 2.8). Make it as tight and compact as you can. The small additional capacitance

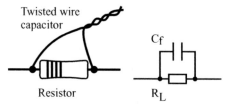

Figure 2.8 The tiny values of extra capacitance often required to stabilize transimpedance amplifiers can be made from tightly twisted wire-wrap wire. Gradually snipping to length allows the transient response to be quickly optimized.

should reduce the receiver's desire to be a generator, and it can be trimmed to the optimum value in a few seconds.

2.5.3   TRY IT! Twisted wire trick

Set up a transimpedance receiver with an LF356 FET opamp and BPW34 or similar photodiode. This has a moderate capacitance of about 72 pF when unbiased. Use a 10-MΩ feedback resistor and attach 50 mm of doubled, twisted 30 AWG (0.25 mm diameter) wire-wrap wire as a small capacitor as shown in Fig. 2.8. Illuminate the detector with light from an LED modulated by a 10-kHz square-wave generator, viewing the receiver's transient response on a scope. Trigger the scope from the signal generator to give a stable display. The transient response should be noticably slower, with a rise time of the order of 100 μs. Now remove 5 mm of the twisted wire capacitance with side-cutters; hopefully this will speed up the response. Take off another 5 mm and look again. With a little experience (i.e., having made two receivers) you will be an expert at this, needing only two or three snips to get the ideal response. If you go too far, the result will be a large overshoot in the transient response, which corresponds to an approach to the resonant condition that caused the oscillation. If you do, make another wire capacitor and try again.

The response obtained in this way may be faster than is needed. Nevertheless, it is recommended to maximize the amplifier's speed as described, even if the eventual application needs a reduced bandwidth. Given that good receivers tend to get used over and over in different applications, it is felt that bandwidth restriction should be carried out in a separate, simple RC low-pass filter external to the amplifier itself. In this way the actual measurement bandwidth is clear in each application and rough estimations of noise can be made. A very useful accessory for photoreceiver development is a set of low-pass filters, wired on small pieces of copperclad board and fitted with terminals and free wires or miniature alligator clips (see Chap. 6). By choosing convenient values, such as a 16-kΩ resistor and 5 percent tolerance capacitors of 1 nF, 10 nF, 100 nF, 1 μF, measurements can quickly be made at detection bandwidths of approximately 10 kHz, 1 kHz, 100 Hz, 10 Hz etc. The standard value from the E24 resistor series is 16 k. This is close enough to the 15.915 that would give decade frequency values. While 15.8 k from the E96 series and precision capacitors will reduce the error such effort is probably wasted on this application.

When building transimpedance amplifiers with high value resistors, say, 10 MΩ and up, often no parallel capacitance is needed, not even a few millimeters of twisted wire-wrap wire. Even without it the circuit is overdamped. This may be because the capacitance of the resistor itself is sufficient. Axial geometry 0.4-W metal-film resistors typically show a terminal capacitance of the order of 0.1 to 0.2 pF. This means that with a 100-MΩ resistor, the impedance of the combination will drop above about 8 to 15 kHz due to this effect alone. Add to this variables such as circuit strays and PCB tracks, and it will be a struggle to get even this bandwidth. This is discussed in more detail later in this chapter.

As we have seen, the transimpedance amplifier can be used with a completely unbiased detector and still give a linear response. Without deliberate bias, the voltage across the photodiode is just the offset voltage of the amplifier. We will assume that this is zero, although even a few millivolts here can significantly affect noise performance, especially if the photodiode has a low shunt resistance.

2.6 Increasing Bandwidth with Loop-Gain

2.6.1 Opamp choice

We have seen that one key to overall photoreceiver detection speed is "gain bandwidth" of the amplifier used in the transimpedance configuration. In an earlier section the bandwidth of a 720-pF/1-GΩ receiver was increased from 0.22 Hz to 363 Hz through the use of a 0.6-MHz GBW opamp, the improvement going approximately as the square root of GBW. Many other devices with higher gain bandwidths are available. For example many common junction FET (JFET) opamp families such as LF353, LF412, TL081, OPA121, AD711 have GBWs around 4 MHz. The OPA604 offers 20 MHz. This should increase our example system's bandwidth to 2 kHz. A few specialist FET amplifiers are available with even greater GBWs. The Burr-Brown (now part of Texas Instruments) OPA657 JFET opamp has GBW = 1.6 GHz, and a unity gain stable device has 230 MHz. In theory this would push our detection bandwidth to 7 kHz.

2.6.2 TRY IT! Opamp choice

Set up a 100-MΩ transimpedance receiver with a small photodiode with about 100-pF capacitance. If this is not available, use a small photodiode with a 100-pF capacitance across it. In a voltage follower configuration we would obtain a detection bandwidth on the order of 16 Hz, or a transient rise time of about 0.35/16 = 22 ms. Use an integrated circuit socket for the opamp. Now take a selection of opamps and observe the variation of transient response for a 5-kHz square-wave input. Figure 2.9 shows my results with a few opamps from the collection. It is clear that the majority of devices from this collection perform similarly, with rise times of 20 to 50 μs, whereas

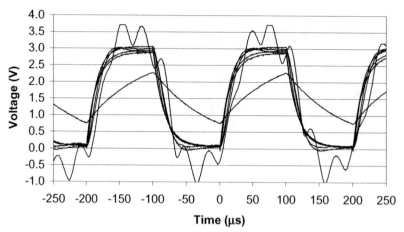

Figure 2.9 Response of a 100-MΩ transimpedance amplifier for a selection of common FET opamps, showing the wildly different responses possible.

a couple show really odd performance. This is not to say that they are poor devices, just unsuitable for the components I connected around them.

2.6.3 Dual amplifiers

The majority of opamps has a DC gain of the order of 80 to 100 dB, or equivalently gain bandwidths of a few megahertz. A convenient and effective approach to increasing GBW well beyond this limit is to combine two or more opamps. The circuit of Fig. 2.10 uses two opamps arranged respectively as a follower with gain and an inverting gain stage. The combination is inverting, just as in a conventional transimpedance amplifier. Ideally, the first device can be chosen for low bias current and low noise, such as the AD711 (GBW = 4 MHz) mentioned earlier, and the second device can be a high-bandwidth bipolar unit, such as an LM833 (GBW = 15 MHz). By choosing the stage gains to be 52 and 192, the respective bandwidths are both equal to 77 kHz. Both amplifiers then act as low-pass filters at 77 kHz, with a combined cutoff frequency of about 54 kHz. Hence the combined amplifier acts like a single amplifier with a gain of 10^4 at 54 kHz. Gain bandwidth then looks like 540 MHz. This should improve the detector time-constant by 23,000 times.

Figure 2.11 is another way to combine two opamps, described in several data sheets from Burr-Brown. Here a composite amplifier is again formed from a low-bias current FET front end and a high-speed bipolar gain stage. No DC local feedback is applied, so at DC the loop gain is the product of the two amplifiers' gains (225 dB). At higher frequencies, the gain is reduced by the integrator around the second amplifier. Transimpedance feedback is applied around both amplifiers, as normally. The signal output is just as for the same feedback resis-

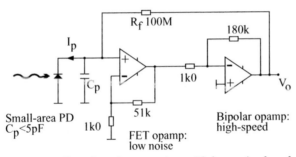

Figure 2.10 Transimpedance receiver with loop gain shared between two different amplifiers. This can improve bandwidth for high parasitic capacitance photodiodes. (*Courtesy of Robert Theobald of Theoptics Ltd.*)

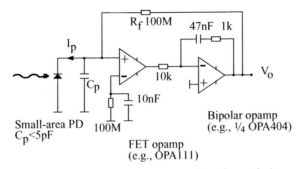

Figure 2.11 Another two-opamp transimpedance design. (*Reproduced by permission of Texas Instruments Burr-Brown.*)

tor with a single amplifier, but signal bandwidth is increased about five times. See the article by Michael Steffes (1997) for further ideas.

2.6.4 Use of discrete components

It appears odd that the opamp, with its complex design and many active elements, offers such a limited GBW. It is not hard to find single bipolar transistors with a high frequency cutoff f_t (equivalent to GBW) of 2 GHz and more, and specialist silicon-germanium transistors used in mobile phone receivers up to 50 GHz. The price paid for stability and ease of use of an opamp with internal frequency compensation is clearly rather high. Hence it is not surprising that discrete transistors can deliver superior designs.

Designs of photoreceivers for fiberoptic communications systems have traditionally been simple circuits with specially built surface-mounted active elements that are wired and constructed on ceramic printed circuits. An example of a generic design is shown in Fig. 2.12. This uses a very low capacitance,

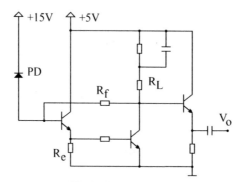

Figure 2.12 The highest speed transimpedance amplifiers are often simple circuits made with heavily reverse-biased, specially selected low capacitance photodiodes and high GBW discrete semiconductors.

heavily reverse biased chip-photodiode connected to an $R_f = 20\,\text{k}$ transimpedance amplifier. The first two transistors in common-collector, common-emitter connections provide the gain, with overall shunt feedback. The last transistor is just an emitter follower to drive the output load. Such circuits are useful up to a few hundred megahertz bandwidth. Even here the stray capacitances and inductances of the particular layout chosen start to dominate performance. Beyond those frequencies, up to the 50 GHz of the current top-end designs, similar principles apply, although the search for special components such as ultra-low-capacitance waveguide-coupled photodetectors becomes a key task. See Umbach (2001) for a discussion of this high-bandwidth detection.

2.7   Big Problem 2: Limitations of the Feedback Resistor

2.7.1   Alternative resistors

In Sec. 2.5.3 we looked at the bandwidth restriction of the parasitic capacitance of a high value resistor. This can often be the dominant limitation. Even with the photodiode represented by a pure, capacitance-free current source and an ideal opamp of infinite gain, system bandwidth can be poor. As we know, the output voltage is just the photocurrent multiplied by the feedback impedance. A standard 0.4-W metal-film resistor typically has an intrinsic capacitance of about 0.2 pF, so if $R_f = 100\,\text{M}\Omega$, the photoreceiver will exhibit a low-pass filtered output with a characteristic frequency $f_c = 8\,\text{kHz}$. In practice, the bandwidth will be even less than this owing to additional stray capacitance from wiring the resistor to its amplifier. One way to improve bandwidth is to find another make or type of resistor with lower parasitic capacitance. In general, the smaller the resistor the better; chip devices perform best.

2.7.2 TRY IT! Resistor choice

Once again, make a transimpedance amplifier based on an FET opamp. Perhaps you should chose the fastest amplifier from the previous TRY IT!. This time let's really go for speed and choose the lowest capacitance photodiode available. Borrow a selection of 100-MΩ resistors. I used a 1-mm^2 UDT PIN040A photodiode with 15-V reverse bias, an OPA604 opamp, and six different 100-MΩ resistors. These range from a 50-mm-long plate construction device used for 500-VAC operation to a standard axial leaded 0.25 W device and a couple of 1 × 2-mm chip resistors.

Illuminate the photodiode with a red LED connected to a square-wave generator, but keep it far enough away not to couple it electrically. This is often visible as greatly enhanced flanks on the received square wave. This is a spurious coupling via the capacitance between LED leads and photodiode leads and gives a high-pass characteristic coupling response. A 100-mm glass or grounded metal tube with an LED and photodiode lodged in each end can be used to increase optical coupling efficiency. Alternatively, a fiber could be used, albeit with a lot more work.

Connect the resistors one at a time and observe the transition rise times. I operated at 5 kHz and used a socketed FET opamp with pins 2 and 6 bent up out of the socket for flying connections. My results are shown in Fig. 2.13a. First we can see the gross differences in performance. The plate resistors show several cycles of overshoot, which can be attributed to the distributed capacitance of these large devices. However even the "normal" components show big differences, with 10 to 90 percent rise time

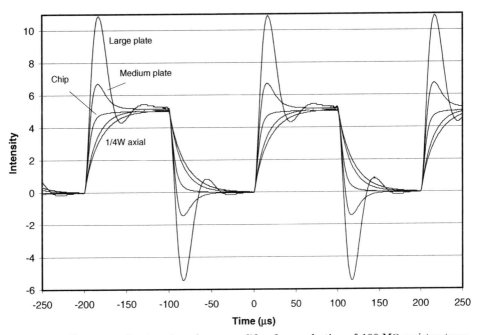

Figure 2.13a Response of a transimpedance amplifier for a selection of 100-MΩ resistor types, including large plate types, 1/4-W axials and a chip types. The fastest response was obtained with a chip component.

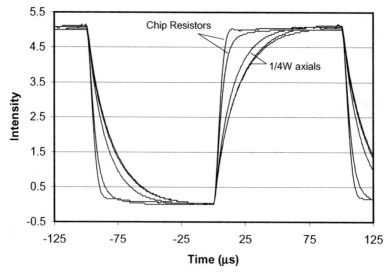

Figure 2.13b Detailed view of the transient response. The slowest responses are for a 1/4-W axial resistor, the two fastest for a chip component wired "casually" and in an optimized compact construction.

varying from 25 to 50 μs. Figure 2.13b looks a little more carefully and adds a further chip device measurement. Here I resoldered the chip as close as I could to opamp pin 2 and generally tidied up the wiring. The result is a further substantial reduction in rise time, down to as short as 5 μs.

All this shows that performance does not require the use of fancy and expensive amplifiers. Equally large gains can be obtained with passive components, in particular the transimpedance resistor, and by careful layout and construction. It also suggests that for all but the simplest of applications, small-package opamps and surface-mount components will give significantly better performance than through-hole designs.

2.7.3 Split resistors

Another, related component choice that is frequently suggested in articles is the use of split transimpedances. Figure 2.14 shows a 100-MΩ, 0.4-W resistor with its intrinsic 0.2 pF parasitic capacitance replaced by two similar 50-MΩ resistors in series. This will almost double the frequency of the -3 dB point. The idea is that although there is double the total capacitance in the feedback, the combination maintains its resistive nature up to higher frequencies. As the output voltage is just the product of the photocurrent I_p with the combined complex impedance of the feedback network, we can easily calculate the output as a function of frequency. With a single resistor the total impedance is:

$$\frac{Z_{C1}R_1}{Z_{C1}+R_1} \qquad \text{with } R_1 = 100 \text{ M}\Omega. \qquad (2.5)$$

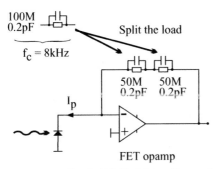

Figure 2.14 Bandwidth-limiting effects of the parasitic capacitance present in all resistors can be reduced by splitting the resistor into two components of half the value.

with split resistors the total impedance is:

$$2\frac{Z_{C1}R_2}{Z_{C1}+R_2} \qquad \text{with } R_2 = 50 \text{ M}\Omega. \qquad (2.6)$$

Figure 2.15 shows the calculated responses, as well as with R_1 split into three and four components. There is indeed a significant improvement in bandwidth from this trick. Figure 2.16 gives the results I obtained experimentally, which are not quite as good as the simulation suggests. Further extension of this idea to synthesize a high-impedance transmission-line resistor in this way looks attractive but becomes progressively less elegant as the number of elements increases.

2.7.4 RC compensation

Another approach to bandwidth improvement, shown in Fig. 2.17, is to compensate the phase and amplitude response of the feedback resistor's high-pass parallel RC combination with a low-pass RC network. In App. A we show how to model this network, which is somewhat more complicated than what we have calculated so far due to the grounded element (C_c). In essence, if $R_fC_f = R_cC_c$, the feedback network looks like a pure resistor, and we should escape the severe bandwidth limitations of C_f.

2.7.5 TRY IT! RC compensation

Set up the transimpedance amplifier as shown in Fig. 2.17 with a 100-MΩ resistor. Use a small-area, low-capacitance photodiode such as the BPX65. Illuminate it using a visible LED and current-limiting resistor driven by a square-wave generator as before. It is best to use an LED with a lensed plastic package, as this allows it to be placed further away from the photodetector while receiving the same output signal.

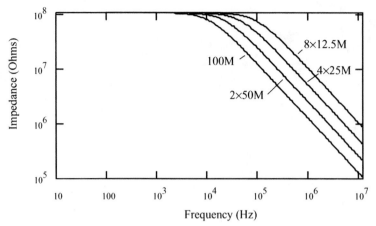

Figure 2.15 Calculations of the frequency-variation of impedance for a 100 MΩ resistor split 2, 4, and 8 ways.

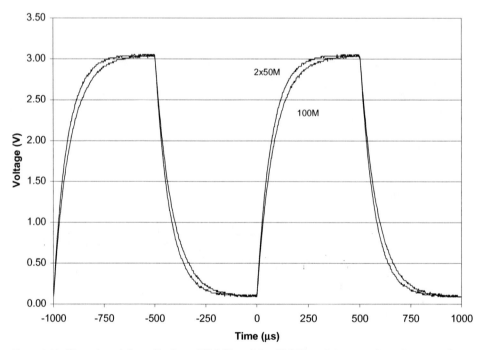

Figure 2.16 Experimental results for a 100-MΩ and 2 × 50-MΩ resistor transimpedance receiver.

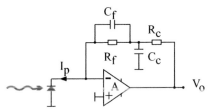

Figure 2.17 The parasitic capacitance may be partially compensated using the additional RC network. Ideally $R_c C_c = R_f C_f$, but make R_c a single-turn trimpot and adjust for optimum.

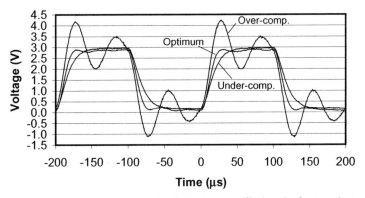

Figure 2.18 Overcompensating leads to severe oscillations in the transient response.

It also allows the use of the glass tube waveguide to improve optical coupling. If it is too close, there may still be interference from *electrical* coupling between LED and photodiode as before. Trigger the scope from the square-wave generator, and observe the transimpedance amplifier output as usual. Adjust the drive frequency and trigger the scope to make the finite rise time visible onscreen. With a BPX65 and OPA604 opamp I measured a rise time (10 to 90 percent) of about $50\,\mu s$.

Now add the RC compensation, using a 20-k trimpot and 1-nF capacitor. See how the transient response varies as the trimmer is adjusted. It is usually possible to speed up the transient response by a factor of two to five times. If you overdo the compensation, serious overshoot or even oscillation will result. My results are shown in Figs. 2.18 and 2.19. You can also check this out in a Spice simulation (Fig. 2.20). It gives the same information but with only half the fun of actually causing the circuit to oscillate!

2.7.6   Shielded resistors

Yet another approach is to try to fabricate a resistor with reduced effective capacitance. This is possible by mounting a conventional resistor in a hole in a

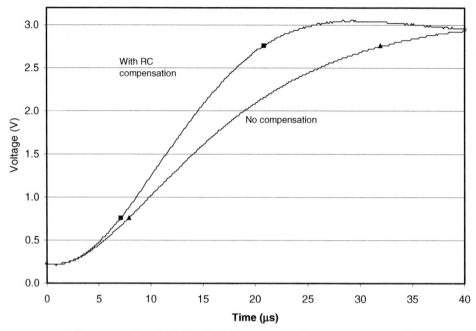

Figure 2.19 RC compensation of a BPX65/100-MΩ transimpedance receiver reduced the 10 to 90 percent rise time by 40 percent, with only slight overshoot.

grounded screen (Fig. 2.21). End-to-end capacitance is reduced in this way, although end-to-shield capacitances may be higher. In my experiments only minimal changes were seen, but in different circumstances performance increases may be greater. With a well-planned PCB layout this may even improve the chip resistor response.

2.7.7   Alternative circuits

The transimpedance configuration is so widely used that it is easy to believe that it is the only way, which would be a pity if it stifles new ideas. In the early days of optical telecommunications systems it was popular to use instead the "integrating" front-end receiver. This was essentially a minimum-size, sub-picofarad pin-photodiode directly coupled to a high-value load (e.g., 5 MΩ) and GaAs FET with selected low gate capacitance, which would nevertheless exhibit a time constant much greater than the bit-period of the incoming digital signal. Without the RC-reduction of the transimpedance configuration, digital data streams were integrated, giving a waveform made up of a number of sawtooth sections. This was then compensated using a CR-differentiator after a stage of amplification. This can give lower noise and better sensitivity than a transimpedance design. The main problem, other than the complexity of high-bandwidth differentiation, is the high dynamic range needed. Long periods without

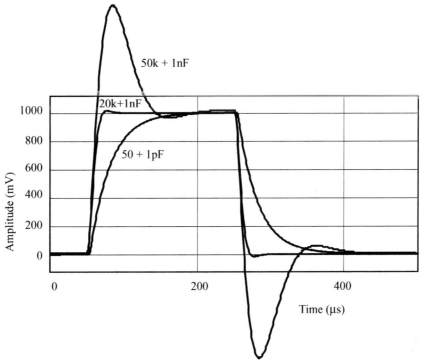

Figure 2.20 Spice simulations of the RC compensation technique show effects similar to those achieved in practice.

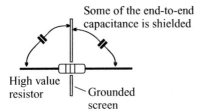

Figure 2.21 Specially customized trans-impedance components may improve bandwidth without any other change to the active components.

data change (strings of 1s or 0s) would cause integration to high voltages and eventual overload. Therefore it is necessary to simultaneously restrict the code words used in the digital signal. This is automatic with selection of return-to-zero (RZ) codes, although non–return-to-zero (NRZ) codes make more efficient use of bandwidth.

Another useful approach to increase bandwidth is to "bootstrap" the photodiode. For example in Fig. 2.6, instead of taking the cathode to ground or to a

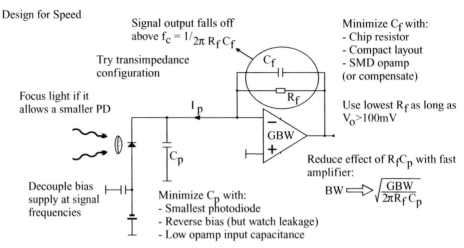

Design for Speed

Signal output falls off
above $f_c = 1/2\pi\,R_f C_f$

Try transimpedance
configuration

Focus light if it
allows a smaller PD

I_p

C_f

R_f

GBW

C_p

Decouple bias
supply at signal
frequencies

Minimize C_p with:
- Smallest photodiode
- Reverse bias (but watch leakage)
- Low opamp input capacitance

Minimize C_f with:
- Chip resistor
- Compact layout
- SMD opamp
  (or compensate)

Use lowest R_f as long as
$V_o > 100$mV

Reduce effect of $R_f C_p$ with fast
amplifier:

$$BW \Longrightarrow \sqrt{\frac{GBW}{2\pi R_f C_p}}$$

Figure 2.22 Summary of the main considerations when designing a transimpedance receiver for speed.

positive bias voltage, it could be driven via a voltage follower with the same voltage that appears on the anode. With no voltage difference, its capacitance is invisible. A single discrete FET or bipolar Darlington transistor source/emitter-follower suffices for this. See Hickman (1995) for a good explanation. Last, big improvements in bandwidth are possible by interposing a discrete transistor common-base amplifier between the photodiode and the transimpedance amp. The photocurrent flows into or out of the emitter, with the collector connected to the inverting opamp input. The base is grounded. As the emitter/collector current transfer ratio is very close to unity, the photocurrent flows also through the transimpedance, as normal. However, the impedance of the emitter is low, reducing the $C_p R_f$ time constant. The transistor should be a small-die radio frequency (RF) type with low collector capacitance, and it must show good current gain at the expected level of photocurrent. This typically restricts the approach to high optical intensities. For a detailed discussion and practical circuits see Hobbs (2000).

2.8 Summary

This chapter has given several circuit fragments for making practical use of your photodiode, and you will find even more approaches in the literature. We have gone through some of the fundamentals and even built a few circuits to see the big differences, especially in detection bandwidth, that small changes in design and even construction can make. The key issues are the capacitance of the photodiode, limiting speed through its interaction with the load resistor, and the difficulty of making large value load resistors which look resistive out to high frequencies. The common 100-MΩ, 0.4-W component has an impedance of $100\,\text{M}\Omega$ only out to $8\,\text{kHz}$. Beyond that its capacitance will dominate performance, and the physical size of the resistor and the details of the circuit layout play a bigger role than the active components. By understanding the principles, we will sometimes be able to make receivers with a bandwidth and sensitivity sufficient for the application. Although it is not really a separate topic at all, but should be handled as if intimately bound with the provision of adequate bandwidth, we put off the calculation of noise to the next chapter.

Fundamental Noise Basics and Calculations

3.1 Introduction

We saw in Chap. 1 that many silicon photodiodes exhibit detection performance in their region of best sensitivity within about 30 percent of the ideal responsivity and limited by fundamental quantum processes $r(A/W) = 0.807\,\lambda(\mu m)$. This *quantum limited* performance does not mean, however, that detectors do everything we want, or even less that detection systems based on them are perfect. The detector might be operating as well as possible but still not well enough. All that really counts for your system is the final measurement signal-to-noise ratio (S/N). We have spent some time looking at the signal (photocurrent), how to maximize it, and how to measure it in a useful bandwidth. Now we must look at the fundamental sources of noise. All detection systems are limited, at least, by shot noise and thermal (or Johnson) noise.

3.2 Shot Noise

Shot noise is the uncertainty in determining the magnitude of a current. It might be thought to be present in any current, whether generated by photodetection or delivered by the wall-socket wiring, although not necessarily with the same relative magnitude. In practice it appears to be relevant only in junctions where there is a "barrier" that carriers must cross, and this includes photocurrents generated in photodiodes. The term *shot noise* arose on listening to the fluctuations in current in vacuum diodes run in their "temperature-limited" region with headphones. Current variations sound like lead shot raining down on a metal plate. If a large number of precision measurements is made of a nominally constant current and the results are plotted, the results should be distributed evenly around the nominal value as shown in Fig. 3.1. On the right of the figure is a histogram that shows the likelihood of measuring a current in a given current "bin." Clearly the most frequently encountered

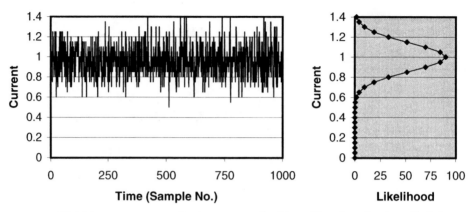

Figure 3.1 Multiple measurements of a photocurrent should provide a Gaussian (actually Poisson) histogram, with a variance equal to its mean.

measured value is the nominal current, here $I_{mean} = 1.0$, but there will always be some fuzziness.

A constant photocurrent (I_p) exhibits a current noise power spectral density (i_n^2) given by:

$$i_n^2 = 2qI_pB \quad \text{or} \quad i_n = \sqrt{2qI_pB} \tag{3.1}$$

where q is the charge on the electron ($1.602 \cdot 10^{-19}$ C) and B is the measurement bandwidth in hertz. This is called the Schottky formula. Note that the current noise power spectral density is proportional to the measurement bandwidth, at least up to some high frequency limit defined by the structure of the photodiode, such as the reciprocal of the transit time of electrons through the detector junction. The spectral density of shot noise is also "white," meaning that the power per unit bandwidth is independent of frequency. The current noise power spectral density can be interpreted as the variance of the current I_p, assuming the distribution of current values looks Gaussian, as in Fig. 3.1. Actually the probability $P(I)$ of a given current cannot be Gaussian, of the form:

$$P(I) \approx e^{-(I - I_{mean})^2 / \text{variance}} \tag{3.2}$$

because this would give a finite probability of our measuring negative currents, which is not physical. The true distribution comes from the analysis of photon statistics.

3.3 Photon Statistics

An alternative description of shot noise in photodetection can be obtained from analysis of photon statistics. When a constant intensity, perfectly monochromatic and perfectly polarized laser source is incident on a photodetector, the actual instant of generation of individual photoelectrons is a stochastic process. For a power of P_s watts, the mean number of photoevents detected per second is $P_s/h\nu = K_m$, where $\nu = c/\lambda$ is the optical frequency, λ is the wavelength, and c is the velocity of light in vacuum. The probability $P(r)$ of measuring r photoevents per second is given by:

$$P(r) = \frac{K_m^r}{r!} e^{-K_m} \tag{3.3}$$

This is the Poisson distribution, which has a variance equal to its mean K_m. The standard deviation is $\sqrt{K_m}$. However, for situations where the mean detection rate K_m of photoelectrons is large, the Poisson distribution becomes identical to the Gaussian distribution. This is why in practice noise current appears to be distributed as a Gaussian function about the mean current, as in Fig. 3.1.

To obtain the full variation in photoevents detected per unit time, the above Poisson distribution for photodetection needs to be multiplied by the stochastic distribution of the incident radiation photons. This depends on whether the incident optical field is highly coherent, like that from a well-controlled laser, or is thermal radiation such as from a hot filament, gas discharge tube, or LED. These complex issues are treated by the subject of photon statistics and are not addressed further here. The interested reader is directed to the excellent book by Goodman and to many other papers on the subject.

For the purposes of this deliberately practical treatment, we can assume that the total photodetection process also exhibits Poisson (or Gaussian) statistics, with variations in photoevents and photoelectrons per unit time exhibiting a variance equal to the mean. This variance in the photocurrent gives rise to the fundamental noise contribution we call shot noise.

The basic results of photon statistics can be applied to the electrons making up the photocurrent from our detector. For example, a photocurrent I_p is made up of a stream of electrons with a mean arrival rate of $I_p/q\,s^{-1}$. In a measurement time of T_i seconds, on average $I_p T_i/q$ electrons will be counted. However, as the electrons follow Poisson statistics, this number can be determined only to within $\sqrt{I_p T_i/q}$. Converting back to current, the uncertainty with which the current can be determined is given by:

$$\frac{q}{T_i}\sqrt{\frac{I_p T_i}{q}} \equiv \sqrt{\frac{q I_p}{T_i}} \tag{3.4}$$

Writing the effective bandwidth of the measurement as $B = 1/(2T_i)$ Hz we obtain for the uncertainty $\sqrt{2qI_pB}$, in agreement with the current noise expression given above.

3.4   What Shot Noise Can We Expect?

Replacing the electronic charge with its numerical value and translating to more useful engineering units we can write for the shot current noise density:

$$i_n = 0.57 \sqrt{I_p(\mu A)} \; pA/\sqrt{Hz} \qquad (3.5)$$

As the shot noise dictates the ultimate precision with which the photocurrent can be measured, it is important to have a feel for its magnitude. Table 3.1 shows a few values.

This shows the precision that is possible in principle if the measurement system is shot noise limited. If we have 1 mA of photocurrent, it should be possible to determine its magnitude with a precision of 1 in 55 million in 1 Hz bandwidth, far beyond the resolution of the best analog-to-digital converters. In practice, many effects conspire to limit our precision to a value much worse that that.

Another big issue is what currents actually show full shot noise? This is a matter of some uncertainty in the literature. Netzer (1981) suggests that shot noise is seen in situations where charge carriers cross a barrier independently of one another, such as pn-junction diodes where the passage occurs by diffusion, a vacuum-tube cathode where electron emission occurs as a result of thermal motion, and photodiodes.

3.4.1   TRY IT! (In)visible shot noise

Instead of looking for parts per million (ppm) variations in milliampere currents, let's try to make an AC optical measurement, with the shot noise as visible as possible. Connect a photodiode and reverse bias it with a 15-V battery, passing the photocurrent directly into a 1-MΩ impedance oscilloscope input (Fig. 3.2). It might be a good idea to carefully filter the battery voltage and put the whole thing in a well-screened metal box. Otherwise, all you will see is line pickup. Illuminate the photodiode with a flashlight or LED to give 1-V DC on the oscilloscope.

This gives us a 1 μA photocurrent, which should exhibit a current noise spectral density of 0.57 pA/√Hz. With the detection bandwidth of 20 MHz, such as that of a modest oscilloscope without bandwidth restriction, the current noise level will be

TABLE 3.1   Some Examples of Shot Noise Values

Current (I_p)	Shot noise current spectral density (i_n)	Relative noise currents $i_n : I_p$ in 1 Hz BW
1 mA	18 pA/√Hz	1 : 55,000,000
1 μA	0.57 pA/√Hz	1 : 1,754,000
1 nA	0.018 pA/√Hz	1 : 55,000
1 pA	0.57 fA/√Hz	1 : 1,754

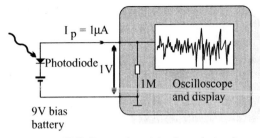

Figure 3.2 (Failed) experiment to show shot noise on the display of an oscilloscope.

about 2.5 nA rms. If this current flows through the 1-MΩ input impedance, the noise voltage will be 2.5 mV root mean squared (rms). The noise on the oscilloscope display with AC coupling should look like a fuzzy band with a width of approximately 6 × 2.5 = 15 mV peak-to-peak (pk-pk). Hence it should be clearly visible and a significant fraction of the 1 V mean voltage.

I tried this and saw nothing! Why not? One of the problems is that the bandwidth of the scope, driven like this from a current source, is not really the 20 MHz you paid for but only a few kilohertz. Yes, it has an input DC impedance of 1 MΩ, but this is shunted by several picofarads. The specification for mine claims 1 MΩ ± 1 percent in parallel with 20 pF ± 2 pF. This means that above a frequency of 7958 Hz, the scope input no longer looks like 1 MΩ, but an ever-decreasing impedance. Put another way, the noise source is driving 1 MΩ with a bandwidth of about 8 kHz; all we can expect onscreen is 0.3 mV pk-pk, which is less than the trace thickness.

How would you increase the level of visible shot noise? You could of course increase the illumination to give 10 V DC at the input. With 10 μA the noise voltage should be increased by $\sqrt{10} = 3.16$ times. This helps a bit, but not enough.

We could drop the resistance to 10 kΩ and increase the illumination further to give the same 1 V output or 100 μA photocurrent. The shot noise should then be 5.7 pA/√Hz, or 0.255mV rms at the scope input. That is less signal voltage than before, although the shot noise current is ten times greater. So we changed in the wrong direction. Upon analysis, for a photovoltage V_{DC} across the input resistor R, the current is $I_p = V_{DC}/R$ and the shot noise current density $0.57\sqrt{I_p(\mu A)}$ pA/√Hz. This can be rearranged to give a shot noise voltage density:

$$v_{\text{shot}} = 0.57\sqrt{(V_{DC}(\text{mV})R(\text{k}\Omega))} \text{ nV}/\sqrt{\text{Hz}} \qquad (3.6)$$

Hence to maximize the detected shot voltage we should have gone for the largest possible resistor with the largest possible DC voltage on it. For instance, with 1 V on 100 MΩ (10 nA) we should have 25 mV rms. Unfortunately, the scope input impedance is too low and doesn't allow for the use of such a large resistive load. Even if we had a 100 MΩ impedance scope, the same capacitance across it as before would limit bandwidth to just 79 Hz, further reducing the total noise power seen on the scope. Increasing the DC voltage is also not so easy. For 100 V we would need a photodiode and scope input that could sustain 100 V or a transimpedance receiver with a 100 V power supply rail built with vacuum tubes or high voltage transistors. All in all, it's surprisingly difficult to demonstrate shot noise in this simple way with a scope; for more convincing demonstrations we will need to design receivers more appropriate than the scope's front-end amplifier. Designing to *see* noise is just as difficult as designing to *minimize* noise. We'll return to this experiment later.

Nevertheless, this experiment points out some of the tricks of design for low noise. The greater the optical power detected and the longer the measurement period, the greater the noise but the less is the relative noise. Hence, other things being equal, we should usually strive in instrument designs for the highest detected power and measure with as small a bandwidth as possible for as long as possible. It also shows that we need to be very careful not to take anything at face value, especially not an innocent looking resistor, even if it is just inside our expensive new scope.

3.5   Thermal (Johnson) Noise

The second fundamental source of noise, thermal noise, was investigated by Johnson and Nyquist in the 1920s. It is present in all resistors at a temperature above absolute zero and is characterized by internal current fluctuations and fluctuations in voltage across their open circuit terminals, even when no external current is flowing. If connected into an external circuit, these will also cause external current fluctuations. Although the warm resistor acts as a little generator, it is not possible to extract power from it. Anything connected to it deposits as much power into the resistor as is extracted from it. Analogous to the treatment of shot noise currents, a resistor of $R\Omega$ will show a noise power spectral density given in voltage or current by:

$$e_n^2 = 4kTBR \text{ (in units of } V^2) \quad \text{or} \quad e_n = \sqrt{4kTR} \text{ (in } V/\sqrt{Hz}) \tag{3.7}$$

$$\text{or} \quad i_{nt}^2 = 4kTB/R \text{ (in units of } A^2) \quad \text{or} \quad i_{nt} = \sqrt{4kT/R} \text{ (in } A/\sqrt{Hz}) \tag{3.7a}$$

where k is Boltzmann's constant $(1.381 \times 10^{-23} \text{J/K})$ and T is the absolute temperature in K (≈ 300 K at room temp.).

Equation 3.7 is the Johnson or Nyquist formula. As with shot noise, this thermal noise power is proportional to the measurement bandwidth. In photoreceiver design it is usually more convenient to calculate and measure voltages and currents rather than powers, so we can rewrite Eq. 3.7 in convenient engineering units, either as a voltage source in series with a (noiseless) resistor R or as a current generator in parallel with R:

$$e_n = \sqrt{4kTR} = 4\sqrt{R(\text{k}\Omega)} \text{ nV}/\sqrt{Hz} \text{ (at room temp.)} \tag{3.8}$$

$$i_{nt} = \sqrt{4kT/R} = 4/\sqrt{R(\text{k}\Omega)} \text{ pA}/\sqrt{Hz} \text{ (at room temp.)} \tag{3.8a}$$

A few calculated values will give a feel for the voltage magnitudes (Table 3.2):

TABLE 3.2 Thermal Noise Voltages

Resistance	Thermal noise voltage density
$50\,\Omega$	$0.89\,\text{nV}/\sqrt{\text{Hz}}$
$1\,\text{k}\Omega$	$4\,\text{nV}/\sqrt{\text{Hz}}$
$1\,\text{M}\Omega$	$126\,\text{nV}/\sqrt{\text{Hz}}$
$1\,\text{G}\Omega$	$4\,\mu\text{V}/\sqrt{\text{Hz}}$

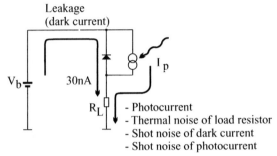

Figure 3.3 Noise contributions of a detector plus bias box.

3.6 Bias Box Noise

It is useful to estimate the detection noise and performance of the simple bias box system described in Chap. 2. We will assume that we are using a BPW34 silicon diode (a long-available plastic encapsulated device with an area of $7.6\,\text{mm}^2$) operated with either $1\,\text{k}\Omega$ or $1\,\text{M}\Omega$ resistors at a $9\,\text{V}$ reverse bias (Fig. 3.3). The noise sources to be considered are the thermal noise of the resistive load, the shot noise of the leakage current, and the shot noise of the signal photocurrent. First consider the load resistance $R_L = 1\,\text{M}\Omega$ exhibiting its thermal noise voltage density of $e_n = 4\sqrt{1000}\,\text{nV}/\sqrt{\text{Hz}} = 126\,\text{nV}/\sqrt{\text{Hz}}$.

Even in total darkness, reverse bias provided by the battery voltage $V_b = 9\,\text{V}$ drives reverse leakage current through the diode. The dark current specified for this diode at $9\,\text{V}$ reverse bias is $30\,\text{nA}$ in worst case, leading to a DC offset on the resistor ($30\,\text{mV}$). It also gives a shot noise current density equal to $i_n = 0.57\sqrt{0.03}\,\text{pA}/\sqrt{\text{Hz}} = 0.098\,\text{pA}/\sqrt{\text{Hz}}$. In the $20\,\text{MHz}$ bandwidth of the oscilloscope this becomes $440\,\text{pA}$ rms. Flowing through the $1\,\text{M}\Omega$ load resistor, this leads to an additional noise voltage of $0.44\,\text{mV}$ rms, or about $2.6\,\text{mV}$ pk-pk. Note that uncorrelated noise contributions must be added as *sums of squares* $\left(V_{\text{total}} = (V_1^2 + V_2^2)^{1/2}\right)$. Table 3.3 summarizes the results, for $1\,\text{K}\Omega$ and $1\,\text{M}\Omega$ loads.

TABLE 3.3 Contributions to Noise in the Bias Box

1 μA Photocurrent 20 MHz Bandwidth	1 kΩ Load	1 MΩ Load
Mean DC signal	1 mV	1 V
Dark DC offset voltage	30 μV	30 mV
Thermal noise of load	4 nV/√Hz or 18 μV rms	126 nV/√Hz or 0.56 mV rms
Shot noise due to 30 nA leakage	0.098 nV/√Hz or 0.44 μV rms	98 nV/√Hz or 0.44 mV rms
Shot noise due to photocurrent	0.57 nV/√Hz or 2.5 μV rms	0.57 μV/√Hz or 2.5 mV rms
Total noise	18.2 μV rms	2.6 mV rms
Signal-to-Noise	55	385

It is clear that a higher value of load resistor gives more thermal noise. However, the larger load also gives a much larger signal, with the result that the S/N improves with the square root of the increasing load. With 1 kΩ the detection S/N is limited by thermal noise in the resistor, the largest of the noise contributions of Table 3.3. With 1 MΩ, however, the S/N is limited by the photocurrent shot noise. The system is "shot-noise limited." The greater the photocurrent, the greater is the noise but the greater the S/N. The only way to remove the signal shot noise contribution is to switch off the optical signal; however, this is not recommended!

It is generally our goal to design the measurement system to be shot-noise limited. This is not to say that the S/N will of necessity be adequate for our measurement, just that we are making best use of the optical power available. From Table 3.1 we can see that if we have high received optical power, then to be shot-noise limited should bring high resolution and perhaps precision. We can easily see whether a simple system is shot-noise limited by measuring the voltage on the load resistor due to photocurrent. To see this, compare shot and thermal noise voltage densities produced by a photocurrent I_p flowing through a simple load resistor R:

$$\text{Thermal noise voltage density} \quad \sqrt{4kTR} \quad \text{V}/\sqrt{\text{Hz}} \quad (3.9)$$

$$\text{Shot noise voltage density} \quad R\sqrt{2qI_p} \quad \text{V}/\sqrt{\text{Hz}} \quad (3.10)$$

Equating these two expressions, we can calculate the product:

$$V_o = I_p R = 2kT/q \quad (3.11)$$

Hence if the signal voltage V_o due to I_p flowing in R is bigger than about $2kT/q = 52$ mV the measurement should be limited by the shot-noise statistics. We say "should" because other interfering signals may actually form the limit to S/N. If $V_o < 52$ mV, thermal noise in the load resistor should define the limit to S/N. This was the justification for not bothering to calculate the load resistor noise contributions in the TRY IT! on shot noise.

Voltage noise density e_n
Current noise density i_n

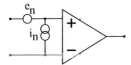

Figure   3.4 Equivalent
noise sources of an operational amplifier.

**TABLE 3.4   Noise Characteristics of Typical Bipolar
and FET Opamps**

Typical bipolar opamp (e.g., OP07)	
Input bias current (i_b)	10 nA
Input offset voltage (V_{os})	0.25 mV
Input voltage noise density (e_n)	12 nV/$\sqrt{\text{Hz}}$
Input current noise density (i_n)	0.2 pA/$\sqrt{\text{Hz}}$
Typical FET opamp (e.g., LMC7101)	
Input bias current (i_b)	64 pA
Input offset voltage (V_{os})	10 mV
Input voltage noise density (e_n)	37 nV/$\sqrt{\text{Hz}}$
Input current noise density (i_n)	1.5 fA/$\sqrt{\text{Hz}}$

3.7   (Operational) Amplifier Equivalent Noise Sources

Whether constructed from discrete components or opamps all amplifier systems
can be described in terms of two frequency-dependent noise sources, a voltage
noise generator e_n and a current noise generator i_n (Fig. 3.4). Both these quantities are usually specified in manufacturers' data sheets for opamps and are
either specified or can be calculated from discrete transistor data sheets. The
current noise generator is essentially just the shot noise of the bias currents
in the amplifier input stages. Depending on bias current there are large differences in current noise spectral densities among opamps, especially between
bipolar devices, which run at relatively high base bias currents, and FET types,
which require much smaller gate bias currents. Table 3.4 compares the noise
generators of a typical bipolar and an FET opamp. Current noise is clearly much
higher in the bipolar device than in the FET device, whereas voltage noise
is somewhat smaller. Nevertheless, both these devices can be labeled "low
noise." Whether one or the other performs better in your measurement is a
function of the circuit configuration. Another important difference to note
between bipolar and FET types is that FET bias currents, and with them
current noise magnitudes, are very dependent on temperature. Hence if high
operating temperatures are expected, these parameters in particular should

be calculated at the temperature extremes. It may be that a conventional or "super-β" bipolar design has smaller bias currents at high temperature than does an FET device.

The equivalent voltage noise is usually due to thermal noise in resistive components, for example the base/emitter resistors of input stage transistors. They are often similar for bipolar and FET opamps, of the order of 5 to $50\,\mathrm{nV}/\sqrt{\mathrm{Hz}}$. However, specialist discrete and integrated devices are available with noise densities down to about $0.75\,\mathrm{nV}/\sqrt{\mathrm{Hz}}$.

Not all noise sources are "white." The current noise and voltage noise parameters which are so necessary to photodetector analysis are therefore not scalar quantities. Inspection of most opamp data sheets indicates that below a certain frequency, called the lower corner frequency f_L, both these noise parameters increase. The frequency variation of noise density is different for FET and bipolar opamps. This is especially marked in the case of the equivalent voltage noise generator. For example, the data sheets for the LMC7101 give $37\,\mathrm{nV}/\sqrt{\mathrm{Hz}}$ at $10\,\mathrm{kHz}$, increasing to $80\,\mathrm{nV}/\sqrt{\mathrm{Hz}}$ at $100\,\mathrm{Hz}$, $200\,\mathrm{nV}/\sqrt{\mathrm{Hz}}$ at $10\,\mathrm{Hz}$, and $600\,\mathrm{nV}/\sqrt{\mathrm{Hz}}$ at $1\,\mathrm{Hz}$. This is assumed to be an instance of so-called $1/f$ noise. Current noise densities also show some $1/f$ character, but usually this is less pronounced and starts at a lower frequency. The $1/f$ character means that the noise power per decade is proportional to the reciprocal of the frequency. The noise varies as:

$$i_n^2 = i_{no}^2\left(1 + \frac{f_L}{f}\right) \tag{3.12}$$

Here i_{no}^2 is the "white" contribution to total noise. It is unclear what is the cause of this universal $1/f$ character, which can be found in an enormous variety of sources and processes; it has been demonstrated in some opamps down to a frequency of $10^{-7}\,\mathrm{Hz}$, an equivalent period of 1 year. The high noise power density at low frequencies strongly suggests that measurements should be carried out at a higher frequency, preferably at audio frequencies or higher. Light sources should therefore be modulated. The advantages of doing this are large, as we will see in Chap. 5.

Either taking only the high-frequency noise spectral densities, or including the full variation with frequency, system noise calculations are "simply" down to determining the output noise contributions from these two noise generators, modified if necessary by any connected circuitry. In practice, this process is sometimes far from simple.

3.8   Discrete Active Component Equivalent Noise Sources

Discrete active components such as bipolar and field effect transistors can be characterized by voltage and current noise spectral densities i_n^2 and e_n^2 in the same way as opamps. The input-equivalent noise sources of bipolar junction transistors are the shot noise of the base current and the thermal noise of the effective base and emitter resistance. These are given by:

$$i_n^2 = 2qI_b = \frac{2qI_e}{\beta} \tag{3.13}$$

$$e_n^2 = 4kT\left(r_b + \frac{r_e}{2}\right) \qquad r_e = \frac{kT}{qI_e} \tag{3.14}$$

I_b and I_e are the transistor base and emitter currents, and r_b, r_e are the base spreading resistance and emitter small-signal resistance. The transistor current gain is represented by β.

For field effect devices the equivalent noise densities are:

$$i_n^2 = 2qI_g + 0.7 \cdot 4\frac{kTC_{gs^2}\omega^2}{g_m} \tag{3.15}$$

$$e_n^2 = 0.7 \cdot 4\frac{kT}{g_m} \tag{3.16}$$

Here g_m is the mutual conductance of the FET, I_g is the gate leakage current, and C_{gs} is the input capacitance of the FET. The operating radian frequency is $\omega = 2\pi f$.

3.9   Design Example: Free-Space Detection of an LED—Big (Resistor) is Better

As another example, let's evaluate the model optical system of Fig. 3.5. This is an infrared TV remote control. Assume that the LED emits $1\,\text{mW}$ at $0.88\,\mu\text{m}$ wavelength, evenly distributed over 2π steradians. At a TV–sofa distance of $2.82\,\text{m}$ the light has expanded to cover half of a sphere of $100\,\text{m}^2$ area. The detector's 7.6-mm^2 active area captures about $150\,\text{pW}$, giving via its $r \approx 0.65\,\text{A/W}$ responsivity a photocurrent of $100\,\text{pA}$. Putting this directly into an oscilloscope's $1\,\text{M}\Omega$ load resistor would give a peak detected voltage of $0.1\,\text{mV}$, which is not visible on the scope. So let's use an amplifier.

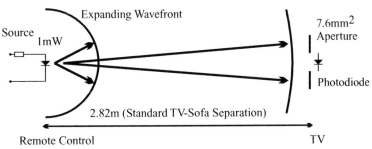

Figure 3.5   An example of an optical communication system: the TV remote control.

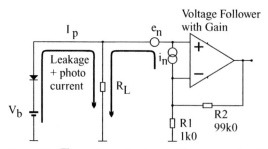

Figure 3.6 The opamp gain stage will increase signal levels but cannot improve the signal-to-noise ratio.

Figure 3.6 shows a simple opamp "follower with gain," using the same $1\,M\Omega$ load resistor as before but a gain of 100 times. We have also included the opamp noise generators. The errors and noise contributions are as follows:

Photocurrent flowing through R_L (signal)

Leakage current flowing through photodiode and R_L (voltage offset)

Opamp bias current flowing through R_L (voltage offset)

Voltage noise generator (output voltage noise)

Current noise generator flowing through R_L (output voltage noise)

Thermal noise of R_L (output voltage noise)

With just $100\,pA$ photocurrent it is clearly not easy to live with the large leakage current of the previous example, so we will assume that this can be kept well below $100\,pA$ by choice of diode and use of only low levels of reverse bias. The signal voltage is then $100 \cdot 100\,pA \cdot 1\,M\Omega = 10\,mV$, just about big enough to be seen on the oscilloscope. However, the load resistor thermal noise is $126\,nV/\sqrt{Hz}$, or $56\,mV$ rms at the output of the opamp. The signal is actually buried under the detection noise, and we would be unlikely to see the signal from the remote LED on the scope.

The $\times 100$ amplifier clearly increases the magnitude of our *signal*, but equally increases the *noise*. The S/N can therefore never be improved in this way. In fact, in the example system the S/N will probably degrade, due to the noise contributions of the amplifier that we haven't even considered yet.

So, how do we improve the detection performance of the LED transceiver? We showed above that the design cannot be shot-noise-limited, as the detected signal voltage is less than $2kT/q$, or $52\,mV$. How about increasing the load resistor R_L to $1\,G\Omega$? With the increase in load resistor, its thermal noise voltage has increased by a factor of $\sqrt{1000} = 32$ times to $4\,\mu V/\sqrt{Hz}$, which seems like a retrograde step. However, the signal from the $100\,pA$ photocurrent has increased by 100 times to $10\,V$. In the $20\,MHz$ bandwidth the thermal noise of the resistor is $18\,mV$ rms. Hence the S/N has been improved by 32 times at a stroke, from $10/56 = 0.2$ to $10\,V/1.8\,V = 5.5$. This is not great, but the signal may now be visible on the scope.

TABLE 3.5 Comparison of Different Load Resistors for Sensitive Detection

Voltage follower: gain = 100× Photocurrent = 100 pA	$R_L = 1\,M\Omega$	$R_L = 1\,G\Omega$
Thermal noise density of R_L	126 nV/√Hz	4 μV/√Hz
Output voltage noise in 20 MHz BW	56 mV rms	1.8 V rms
Output voltage noise in 100 Hz BW	0.126 mV rms	4 mV rms
Output signal voltage	10 mV rms	10 mV rms
S/N in 100 Hz BW	79	2500

TABLE 3.6 Comparison of Different Amplifier Types

Voltage follower: gain = 1× Photocurrent = 100 pA $R_L = 1\,G\Omega$, 100 Hz detection BW	Typical bipolar opamp (e.g., OP07)	Typical FET opamp (e.g., LMC7101)
Input bias current (i_b)	10 nA	64 pA
Input offset voltage (V_{os})	0.25 mV	10 mV
Input noise voltage density (e_n)	12 nV/√Hz	37 nV/√Hz
Input noise current density (i_n)	0.2 pA/√Hz	1.5 fA/√Hz
Signal output	0.1 V	0.1 V
Static offset voltage due to V_{os}	0.25 mV	10 mV
Offset voltage due to i_b	10 V	64 mV
Amplifier voltage noise in BW	120 nV rms	370 nV rms
Amplifier current noise in BW	2 mV rms	15 μV rms
Load resistor thermal noise in BW	40 μV rms	40 μV rms
S/N in BW	50	2300

The next thing we can do is to realize that the TV remote control needs a signaling bandwidth only of the order of a few tens of hertz. Hence let's additionally restrict the bandwidth by adding a low-pass filter with an upper cutoff frequency of 100 Hz. The 1 GΩ load thermal noise voltage drops further to 4 mV rms. Suddenly we have a S/N ≈ 2500, and the system is starting to look useful. Table 3.5 summarizes the calculations.

Having analyzed the S/N for a noiseless amplifier, we can go on to add the opamp voltage and current noise generators to our 1 GΩ design. This is shown in Table 3.6.

It is clear that in this application the use of the FET opamp is preferable due to its low bias current. The bipolar opamp's high bias current leads to a large static offset voltage. The FET opamp greatly reduces this effect, as well as the related noise contribution from its current noise generator. Nevertheless it would be useful to reduce the 64-mV output offset voltage due to bias current, for example by using an amplifier with even smaller bias current.

3.10 TRY IT! Another Attempt at Visible Shot Noise Measurements

Let's look again at noise onscreen. This time we'll increase the load resistor and add some gain to make it more detectable with an oscilloscope. Connect a small-area

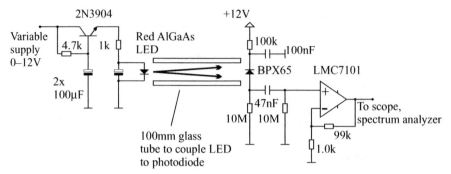

Figure 3.7 Another shot-noise demonstrator. A quiet current source drives the LED. An optical waveguide increases optical coupling efficiency without increasing electrical interference coupling.

photodiode such as a BPX65 to a 12V reverse bias supply and 10MΩ load resistor. We want to have a few volts across the load, so AC couple to remove the DC signal before amplification. Something like Fig. 3.7 should suffice, with a follower-with-gain set up for 100× gain. For the photocurrent generation I again used a high-brightness red (670 nm) AlGaAs LED pushed into a 100 mm length of 5 mm ID glass tube wrapped in black tape. With the photodiode in the other end we get good optical coupling, with very little electrical coupling. For decent noise measurements it is important to remove all the sources of interference, and an LED on long drive leads is one risky area. You could use imaging optics or a length of fiber, but I find the tube is more convenient. When using a power supply it is also preferable to filter the LED drive current as we're going to be able to see rather tiny variations in LED output and don't want current variations disturbing our measurement. The transistor capacitance multiplier does a good job of smoothing the LED current. With a battery you may get by without it.

Looking at the opamp output on a scope, you should see a trace with a millivolt or so of noise, but you don't want to see big line voltage waveforms. If you see any that sit stably on screen when the scope is triggered from the AC line voltage, try to improve the grounded metal screening around the circuit or check for pickup on the various leads. With a (>10 MΩ impedance) voltmeter on the resistor load, wind up the LED current to get 1 V DC. The photocurrent is only 0.1 µA, but this is a good place to start. Don't worry about the scope display at this stage, it will probably be erratic because of noise being injected from the voltmeter. Remove the voltmeter again and have another look. You should be able to see a distinct increase in peak-to-peak noise as the LED brightness is increased. I measured about 30 mV pk-pk.

If you have a spectrum analyzer, have a look at the low-frequency noise for various photocurrents. I measured −71 dBm in 5.6 Hz bandwidth with the LED off and −58 dBm with it on. We can estimate the shot noise signal of the 0.1 µA current as $0.18 \, pA/\sqrt{Hz}$, or 0.2 mV rms in the measurement bandwidth after amplification (−60 dBm). Without the LED illumination we should just have the thermal noise of the 5 M parallel combination of resistors. This is 67 µV or −70.5 dBm. This is a reasonable agreement.

Just for fun, replace the photodiode and its 100 kΩ resistor with a forward-biased silicon diode such as a 1N4148 plus another 10 MΩ resistor. Reduce the bias to still have the same current in the load resistor as before. I measured a noise level indistinguishable from that of the resistive load without any current flowing. This suggests that the forward-biased silicon diode carrying current does not show full shot noise. Maybe it does, but the externally measured noise is modified by the low impedance of the forward-biased diode. We will come back later to noise measurement in forward-biased diodes. Neither do the current-carrying resistors seem to show shot noise. Netzer (1981) states that metallic conductors carrying current do not show shot noise "because of long-range correlation between charge carriers." Hobbs (2000), too, confirms that current-carrying resistors show a level of noise far below shot level.

3.11 Dynamic Noise Performance

3.11.1 Amplifier dynamic noise calculations

Up to now we have considered only static noise sources and modification by the connected amplifier. However, as we have seen, feedback resistors exhibit stray capacitance, photodiodes have parasitic capacitance, and amplifiers have input capacitance and frequency-dependent gain characteristics, all of which combine to modify the overall noise as a function of frequency. In this section we address the main features which can significantly affect dynamic system performance. We will restrict the discussion to circuits with opamps, but the principles are equally applicable to any amplifier design. The approach is simply to apply well-known feedback amplifier gain expressions, including all the significant frequency-dependent components.

The follower configuration of Fig. 3.8 is straightforward, as the load seen by the photodiode current generator is the parallel combination of the load resistor R_L, its parasitic capacitance C_s, the input capacitance of the amplifier C_i, and the parasitic capacitance of the photodiode C_p. In most cases C_s will be negligible in comparison with the others. We will lump all the contributions in with C_p. Writing for the total frequency-dependent impedance:

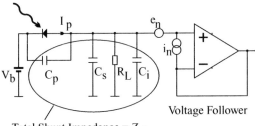

Figure 3.8 In a bias box or voltage follower configuration, noise can be estimated by placing all stray and input capacitances in parallel with the photodiode current generator.

$$Z_{sh} = \frac{R_L Z_{Cp}}{R_L + Z_{Cp}} \qquad\qquad (3.17)$$

$Z_{Cp} = 1/sC_p$ is the impedance of the total parasitic capacitance, and $s = j2\pi f$ is the complex angular frequency. Appendix A gives some pointers on the use of the complex frequency to calculate such electrical networks. The signal output is just $I_p Z_{sh}$. (This signal shows a low-pass characteristic, starting at $I_p R_L$ at a low frequency, before reducing above the break frequency $1/2\pi R_L C_p$.)

The amplifier's current noise generator flows similarly through the shunt impedance Z_{sh}, so that both photocurrent and current noise contributions at the amplifier output exhibit the same low-pass characteristic. The voltage noise density generator's output simply appears at the amplifier output and is essentially independent of frequency. Above some frequency it will therefore dominate the noise density. If the voltage noise contribution is negligible, then S/N is constant with frequency.

The easiest way to treat the thermal noise of the load resistor is to consider it as another current source $4/\sqrt{R(k\Omega)}$ pA$/\sqrt{\text{Hz}}$ in parallel with the photocurrent and shot noise sources. All three current sources then show the same low-pass characteristic due to the falling shunt impedance Z_{sh}.

3.11.2 Transimpedance amplifier: noise peaking

For the transimpedance configuration, the situation is less straightforward (Fig. 3.9). As before we lump together all elements of the photodiode equivalent circuit, input parasitics, and those of the feedback elements into general input and feedback impedances Z_{sh} and Z_f, respectively. If the opamp is ideal (zero bias current and infinite gain), both signal photocurrent and current noise flow through the feedback impedance Z_f, which is the parallel combination of the load resistor R_f and its parasitic capacitance C_f in the figure. As in

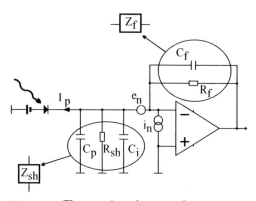

Figure 3.9 The transimpedance configuration can be analyzed using the total input and feedback impedances.

the follower configuration, above a break frequency f_1 the signal output decreases at −20 dB/decade. Writing in the complex frequency (s) domain the impedance of C_f:

$$Z_{Cf} = \frac{1}{sCf} \tag{3.18}$$

$$Z_f = \frac{R_f Z_{Cf}}{R_f + Z_{Cf}} \quad \text{(parallel combination of } Z_{Cf} \text{ and } R_f) \tag{3.19}$$

$$f_1 = \frac{1}{2\pi R_f C_f} \tag{3.20}$$

That is the photocurrent, feedback resistor thermal current noise, and amplifier current noise outputs decrease together above f_1.

The effect of the circuit configuration on the amplifier's voltage noise is more complicated. The voltage noise generator sees an inverting amplifier with a gain given by the ratio of feedback to input impedances. Hence this noise is amplified to:

$$e_n\left(1 + \frac{Z_f}{Z_{sh}}\right) \tag{3.21}$$

where: $Z_{sh} = \dfrac{R_{sh} Z_{Cp}}{R_{sh} + Z_{Cp}}$ (parallel combination of C_p and R_{sh})

C_p as before contains all the input capacitances. This contribution to noise therefore *increases* at 20dB/decade above a break frequency f_2 where the impedances Z_f and Z_{sh} are equal. With large transimpedance resistors and/or high-capacitance photodiodes this will be typically $f_2 = 1/2\pi R_f C_p$. The various characteristic frequencies are shown in Fig. 3.10 as a schematic Bode plot. The numerical values shown assume an $R_f = 100\,\text{M}\Omega$ transimpedance, $C_f = 1.6\,\text{pF}$ feedback parasitic capacitance, and a $C_p = 160\,\text{pF}$ detector capacitance.

Where large detectors are used with their high capacitance and consequently low impedance, and high value transimpedances, the magnification of the voltage noise caused by this effect can easily become the dominant noise source. As can be seen from Fig. 3.10, the noise gain increase is checked by the parasitic capacitance of R_f and eventually reduced by the dropping open loop gain of the amplifier. In between a region of high excess noise is often seen, which is termed *gain peaking*. Adding extra capacitance across R_f can lower the f_1 break frequency to reduce the peak, but only at the expense of signal bandwidth.

Given a particular photodiode, large enough to collect most of the signal light, and a transimpedance to provide adequate signal voltage, there isn't much we can do about gain peaking except look for amplifiers that contribute the lowest possible noise. It is one of the surprising aspects that despite working with nA

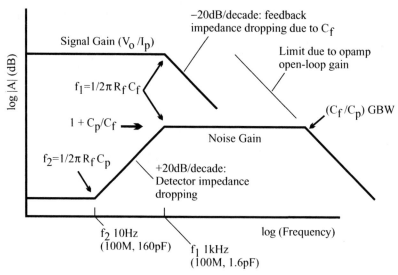

Figure 3.10 Noise peaking. Signal gain is limited primarily by the transimpedance capacitance C_f. The amplified voltage noise e_n increases with frequency and can dominate the noise power for high-capacitance photodiodes with high-value transimpedance resistors.

photocurrents and $100\,M\Omega$ resistors, it is often not the current noise but the voltage noise density e_n that defines our overall performance. The majority of IC opamps exhibits an $e_n \approx 20\,nV/\sqrt{Hz}$ in the flat region above the onset of $1/f$ noise, but there are exceptions. The Burr-Brown OPA121 is an FET-input 2 MHz gain-bandwidth device with 5 pA bias current and noise down to $6\,nV/\sqrt{Hz}$. The Linear Technology LT1792/93 offer bias currents down to 10 pA and e_n typically $4.5\,nV/\sqrt{Hz}$. This is about the limit now for integrated FET amplifiers.

3.11.3 Nanowatt detection calculations

Let's analyze a real example in this way. This is a sensitive receiver for 1 nW detection at low audio frequencies. A $100\,M\Omega$ transimpedance has been chosen, which exhibits a parasitic capacitance of 0.15 pF. This is increased to 1.6 pF by circuit board strays. The photodiode is a $10\,mm^2$ device with a capacitance of 160 pF. Table 3.7 summarizes the calculated results and Fig. 3.11 shows the calculated frequency responses. You can see that although at 10 Hz the amplifier current noise density is greater than the voltage noise density contribution, there is a crossover at about 100 Hz. Up to 1 kHz the transimpedance thermal noise is greater than both amplifier noise contributions, but just less than the signal shot noise for this 1 nA photocurrent. Above about 1.5 kHz the receiver is no longer shot-noise limited, due to the increasing amplifier voltage noise gain peaking.

The design highlights the severe limitations of even small amounts of stray capacitance on the high-value feedback resistance. Even 1.6 pF causes a drop in

TABLE 3.7 Transimpedance Analysis of a 1nA Receiver

Transimpedance amplifier (FET type)	
Signal photocurrent I_p	1 nA
Feedback resistance R_f	100 MΩ
Feedback capacitance, including wiring C_f	1.6 pF
Photodiode capacitance C_p	160 pF
Input noise voltage density (e_n)	10 nV/√Hz
Input noise current density (i_n)	1.1 fA/√Hz
Signal, current noise break frequency (f_1)	1000 Hz
Voltage noise break frequency (f_2)	10 Hz
Thermal noise of feedback resistor at 1kHz	0.63 μV/√Hz
Current noise contribution at 1kHz	0.05 μV/√Hz
Voltage noise contribution at 1kHz	0.5 μV/√Hz
Signal shot noise voltage	0.9 μV/√Hz

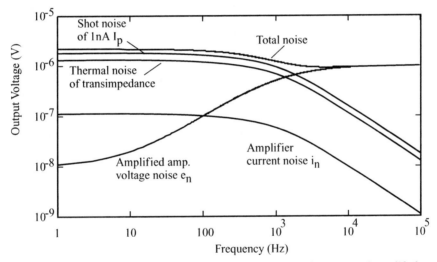

Figure 3.11 Calculation of noise contributions, showing the dominance of amplified e_n above about 1500 Hz.

signal gain above 1 kHz. Equally, the 160 pF photodiode capacitance causes voltage noise to be magnified starting at 10 Hz. Between this frequency and the second break, this noise peaks by almost 40 dB.

3.11.4 TRY IT! Single-mode fibers with visible LEDs

As an illustration of what you might do with a nanowatt receiver, let's look at the use of single-mode fibers, not with laser sources but with LEDs. You can use the small light source for a high-resolution scanner or make use of the wide availability of fiber connectors, "connectorized" detectors, power splitters, and optical bench hardware to ease the construction. Perhaps you want to do measurements of their birefringence properties at a wide range of wavelengths, and a selection of LEDs is much cheaper than a tunable laser. Take an ordinary visible-wavelength LED, say a bright blue one (470 nm), and remove the molded lens to get down to within a few hundred microns of the chip surface (see Chap. 6 for instructions). Drive the LED at its rated peak current with a square-wave generator at 3 kHz. Take one of your 100-MΩ receivers designed using the above principles, and make sure it is totally light-tight in a grounded metal box. Check too that it is working properly. Waft the modulated LED near to the slightly open lid, and you will see nice square waves and plenty of receiver bandwidth register on the oscilloscope.

Let's make life easy for now and use a standard telecomms-type fiber (e.g., a few meters of Corning SMF28, but anything similar will do). Couple the cleaved fiber end to the LED. You could use an XYZ micromanipulator, but with that big chip adhesive tape holding the fiber and a few bending adjustments of the LED leads should couple well enough. There is even room for several fibers. Most important: switch the lights off or put a blackout sheet over your head, and make sure that blue light is coming out of the fiber end. Put the fiber at about 250 mm from your eye. If you can see the light, you should have no difficulty measuring it.

Couple the other end of the fiber to the photodiode. Again, sticky tape and judicious bending will work, although a precision V-groove fiber holder or connectorized detector and "bare-fiber coupler" make life easier still. Observe the oscilloscope. I got about 200 mV from the 100 MΩ transimpedance (2 nA or about 8 nW). On the spectrum analyzer, I measured a signal of –6 dBm and a noise level in the vicinity of 3 kHz of –88 dBm in 45 Hz bandwidth. So a S/N of 82 dB in this bandwidth, useless for telecomms but ample for a lot of laboratory and instrumentation measurements. Note that the receiver is not optimized for lowest noise. It is clear that you don't necessarily have to restrict yourself to lasers or even to the design wavelengths of single-mode fibers. There will be a few modes there, but if all you want is a small light source, this may suffice. If you want to take this experiment further, you can do the same with a single-mode fiber designed for 470 nm. Their core area is another order of magnitude smaller, but even with 0.1 nA photocurrent high resolution, low bandwidth measurements are possible.

Figure 3.12 shows the receiver noise level from DC to 20 kHz in 180 Hz bandwidth. The photodiode used (capacitance 16 pF) showed only mild noise gain peaking at the modulation frequency, so I exaggerated it with 100 pF and 330 pF capacitors, connected across the photodiode. This gave a marked effect. With 330 pF there is about 30 dB of peaking rising up to 5 kHz. At low frequencies (<1 kHz) the noise level is similar to before, that is, limited by the resistor thermal noise. If the detector capacitance of 330 pF is unavoidable, the best S/N will be obtained at frequencies below 1 kHz, although measurement at higher frequencies may be preferable to avoid interferences. Clearly the final choice will be a compromise, reached after detailed S/N measurements in the range of interest. Note too that the amplifier type can make a big difference in noise performance, which is not necessarily predictable from the abridged data often available. Figure 3.13 shows the 100 MΩ receiver noise with a selection of opamps. There are >30 dB differences visible even at these modest audio frequencies. The large differences are probably due to differences in opamp input capacitance, as well as voltage noise density.

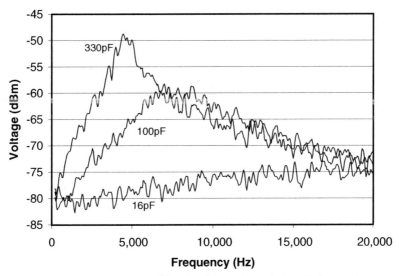

Figure 3.12 Measured noise peaking for three values of photodiode capacitance.

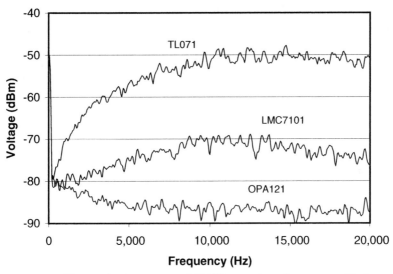

Figure 3.13 Measured noise spectra of $100\,\text{M}\Omega$ transimpedance amp with three opamp types.

3.11.5 Discrete front ends

When we need to use a physically large and therefore high-capacitance photo-detector, this is one situation where a discrete transistor front end can be very useful. Here the dominant noise source at anything above a few hertz is

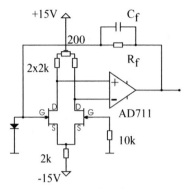

Figure 3.14 Use of a discrete FET with an opamp to reduce the voltage noise density e_n. This is especially useful with large, high-capacitance photodiodes.

going to be the multiplied voltage noise density of the amplifier. Consequently it pays to find a device, typically an FET, with the lowest possible voltage noise. The lowest voltage noise FETs are usually obtained with large dice, for example National Semiconductor's Processes 51 (J113) and 92 (J309 and similar) or the Toshiba meshed gate devices (2SK147 and similar). These can approach $1\,\mathrm{nV}/\sqrt{\mathrm{Hz}}$, which is significantly better than most opamps. The downside is high gate-source capacitance. This is not usually a problem since the photodetector already has a large capacitance and is the reason for adding this FET. There are many ways to use FETs with opamps. The usual approach is to use a pair of matched FETs as a differential amplifier in a "long-tailed-pair" configuration driving an opamp, with overall feedback (Fig. 3.14; Maxwell 1982; Pease 2001).

3.12   Noise and Signal-to-Noise Measurement

3.12.1   Noise measurement tools

Noise measurement systems are as difficult to design as the low-noise electronics being tested and for all the same reasons. The test electronics used must be designed for lowest noise so as not to swamp the noise you are looking for. Frequency-dependent elements must be understood and characterized to avoid serious modifications to the noise spectra. Interferences from line voltages, monitor refresh signals, and clock signals must be suppressed to insignificant levels. Despite the difficulties, it is important for almost every project to have a go and determine the S/N, understand the limiting factors, and dig further into any unexpected aspects. Several tools can be put into service for this.

3.12.2   Use of an oscilloscope

The first job is always to have a good look on the oscilloscope. It is amazing how often one sees detectors plugged straight into instrumentation and PC loggers, without any idea of what the signal is like. This is asking for problems. With a scope and the sensitivity turned up so that you can see the variations, you will be able to get some idea of the magnitude and frequency content of the noise. Gaussian noise has a certain "look," and it is often possible to detect variations from the correct one. It should for instance be bunched around a mean value, be symmetrical, and have the right fuzziness. A problem is that you can get almost any fuzziness you like by changing the scope time base. Wind it up to high speed and the trace might be a straight line. Slow it down and the noise will just look like a thick band. It is therefore important to have an idea of the frequency band that the noise occupies. If white noise is band-limited at 1 Hz and displayed with 1 s/cm on a scope or chart recorder, it should look the same as noise band-limited at 10 kHz, and displayed at 100 μs/cm. The look is defined by the ratio of time-base (or spreadsheet X-axis scale) to noise bandwidth.

The problem is compounded these days with the dwindling availability of analog scopes. The analog scope usually draws many traces during the persistence time of the phosphor screen. Faster time-bases give more overlaid traces, which the screen and the eye tend to average. The digital scope, and especially that with a digital LCD screen, erases the previous trace before drawing a new one and typically makes only a few updates per second. The averaging of a persistent phosphor is lost. We still have some averaging (about 100 ms) due to the eye's response time, but the difference in look can be drastic. Neither is right or wrong, but it's best to have both types of scope available.

Even with these difficulties, the scope can be used to estimate noise amplitudes. If it is Gaussian, then the displayed voltage should spend 99.7 percent of its time within $\pm 3\sigma$ of the mean (σ is the standard deviation). We can take the visible peak-to-peak voltage as 6σ. Of course, if it is Gaussian, you will always see a bigger voltage if you wait long enough, so the peak-to-peak value should not be taken as a precision estimate. It is said that *two* displayed noisy waveforms, aligned and offset on the scope to fit together without visible join, can be used to improve precision. In practice it is probably easier to just take a guess at the peak-to-peak.

The scope is also useful to detect non-Gaussian character. Popcorn noise from poorly passivated photodiodes and uncleaned PCBs can give a waveform with a large number of steplike jumps. Asymmetry can suggest problems with dynamic range, for example with a DC voltage approaching the supply rails.

3.12.3   Spectrum analyzer

After the scope the electrical spectrum analyzer is the most powerful instrument for noise measurements. With its hundreds of frequency-resolved voltage measurements per scan, it is usually easy to separate spot-frequency interference from white noise signals, see the effects of bandwidth limitation, and make quantitative measurements over a wide dynamic range, as long as the signals are quasistatic. Unfortunately, the fabulous performance of a good spectrum analyzer does not come cheap, so they are found in labs only about 1 percent as often as a scope. As discussed in App. II, software-based frequency analysis will suffice, but requires great care.

3.12.4 Narrow band filters

If you do a lot of noise measurements, a cheap but fairly good alternative to the spectrum analyzer is a narrow band filter (Fig. 3.15). By measuring over a known and restricted frequency band, the main interfering sources can be avoided. Frequencies of 2 to 20 kHz seem about right for many instrument designs and their modulation and synchronous detection frequencies. This is well above most industrial noise but not so high that every parasitic capacitance affects the power density.

The filter can be a passive LCR resonant circuit as described in Chap. 6 or an active opamp design (Fig. 3.16), as long as the signal is large to enough to dominate the noise. The center frequency here is given by (Berlin 1977):

$$f_o = \frac{1}{2\pi}\left[\frac{1}{R_5 C_2 C_4}\left(\frac{1}{R_1}+\frac{1}{R_3}\right)\right]^{1/2} \tag{3.22}$$

The passband gain is:

$$G = \frac{R_5}{R_1(1+C_4/C_2)} \tag{3.23}$$

Usually the two capacitors are chosen equal, when we can write simplified expressions in the quality factor (Q):

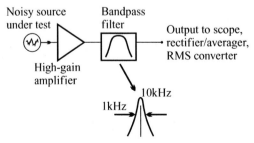

Noisy source under test

High-gain amplifier

Bandpass filter

Output to scope, rectifier/averager, RMS converter

1kHz   10kHz

Figure 3.15 Narrow-band filters are useful to estimate the noise at one spot-frequency. The quasi-sinusoidal output can be quantified using a scope or AC voltmeter.

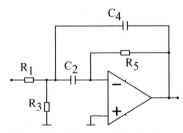

Figure 3.16 Opamp active bandpass filter.

$$R_1 = \frac{Q}{GC2\pi f_o} \quad R_3 = \frac{Q}{(2Q^2 - G)C2\pi f_o} \quad R_5 = \frac{Q}{C\pi f_o} \quad G = \frac{R_5}{2R_1} \quad (3.24)$$

Even looking at the output of the filter on a scope can be useful. With a Q of 10, the filtered noise will look pretty much like a sine wave with variable amplitude, and this peak-to-peak amplitude can be estimated as before or quantified using an operational rectifier circuit (Fig. 3.17). Rms noise powers can be measured accurately using a thermistor bridge (Fig. 3.18). After balancing the bridge for zero differential output, the noise signal is AC coupled to the thermistor, which heats it up. The bridge imbalance is then measured from the instrumentation amplifier output. Because the temperature rise is proportional to the noise power injected, the rms amplitude is determined substantially independent of the shape of the heating waveform. This is convenient since the detector can be calibrated using a sine wave of known amplitude. Alternatively, you might use an integrated circuit rms converter to obtain the average amplitude of the filtered noise. These are normally specified by their *crest factor*, which is the peak-to-mean ratio they can handle with a given error. Electronic systems are limited by supply voltages and generally do not perform as well as the thermistor thermal detector with large crest-factor signals. An rms-computing IC such as an Analog Devices AD636 can typically handle a crest factor of 6 for 0.5 percent additional error.

3.12.5 Lock-in amplifiers

Lock-in amplifiers are also useful for noise measurements. One technique for generating low-frequency white noise, free from any spectral "color" due to $1/f$ noise, is to generate noise at high frequency, say 100 kHz. At this frequency the source spectrum is flat. Next frequency-shift down to DC using the multiplier of the lock-in driven at 100 kHz (Fig. 3.19). The lock-in, being a high-performance frequency-selective vector voltmeter in its own right, can also be thought of as a single-channel spectrum analyzer. By driving the reference input

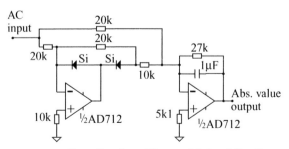

Figure 3.17 Operational rectifier useful for AC voltage measurements of noise. Adapted from National Semiconductor Corporation, "Linear Applications Databook," 1986.

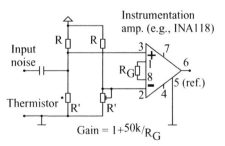

Figure 3.18 Waveform-independent noise power measurements can be made using heating of a thermistor to unbalance an electrical bridge.

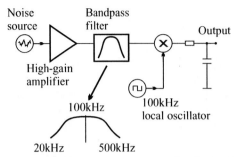

Figure 3.19 Flat noise sources can be obtained by frequency shifting high-frequency noise using the multiplier of a lock-in amplifier.

TABLE 3.8 Noise Bandwidths and Rise Times for Two Orders of Lock-in Postdemodulation Filters

Filter type	6 dB/octave	12 dB/octave
ENBW	1/4 TC	1/8 TC
Rise time (10–90%)	2.2 TC	3.3 TC
Rise time (0–95%)	3.0 TC	4.8 TC

with the frequency of interest and setting the detection bandwidth, noise measurements can be made very conveniently. It is best to use the magnitude/phase angle mode (Rθ) mode, which automatically computes the magnitude of each spectral component. Set the reference frequency to the region where you want the noise measured and adjust the bandwidth, usually to a rather narrow value. The amount of noise passed in the bandwidth defined by the postdemodulation filter time constant depends on the details of the filter used. This is called the equivalent noise bandwidth (ENBW). Table 3.8 gives the values for my two-phase lock-in.

Some commercial lock-ins include a "noise measurement" facility. This is something like an Rθ mode measurement with a very long time-constant (e.g., 10 s) and with AC-coupling to the detector.

3.12.6 Noise generators

It seems perverse that noise *generators* can be useful in the struggle to reduce noise in detector designs. Nevertheless, they are sometimes helpful, if only to test the performance of filters and our noise measurement kit. For our needs, we generally look for a generator that provides white noise, without any marked frequency characteristic, and has a constant mean power per unit bandwidth. This is not the case in some fields, such as testing audio systems. Then it is sometimes more useful to use "pink noise," with a constant power per octave of bandwidth, or other distributions. Here the power per unit bandwidth rises at 10 dB/decade toward lower frequencies.

For highly repeatable, calibrated measurements, it is still of interest to use a thermionic (vacuum) diode operated under so-called temperature-limited conditions, where current through the diode changes only slowly with voltage, and there is no space charge. However, the relatively high voltages and the filament heaters needed for valve operation don't fit in well with current electronic equipment.

In principle, any semiconductor device or even resistor can be used as a generator, with sufficient amplification, using their shot or Johnson noise spectral densities. The difficulty is in ensuring a flat frequency spectrum. Devices that produce much higher amplitudes are useful to reduce the problems of $1/f$ noise in the amplifier. The zener diode is often used for this task. It may be better to use high-voltage types because above breakdown values of about 5.7 V Zener diodes operate via avalanche processes and have a positive temperature coefficient. Below this value they operate with the zener effect and have a negative temperature coefficient. Any ordinary semiconductor device operated in reverse breakdown mode may also work well. I have not been able to find in the literature any consensus on this. Published circuits typically use a 12 V zener diode, operated at a current of the order of 1mA with AC coupling to a low-noise amplifier. I tried instead a small npn silicon transistor (ZTX453), with the base-emitter junction reverse biased to breakdown, operated into a transimpedance amp (Fig. 3.20). Below the breakdown voltage of the junction, the noise will be just that of the amplifier and its load resistor. Just at breakdown, about 8.30 V in my example, the noise level increased to an enormous value. In fact it is not even necessary to use the transimpedance amp. With a 100 k load the scope input is sufficient to see the noise. In the time-domain the waveform looks like a slow sawtooth with fast jumps between two voltage levels and stable stretches of many tens of microseconds. As the bias current is increased, the slope of the sawtooth increases, putting more energy into higher frequencies (Figs. 3.21,

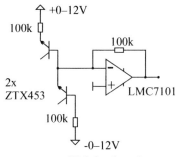

Figure 3.20 High-level noise can be conveniently generated using a bipolar transistor base-emitter junction biased to breakdown. The use of two devices can avoid AC-coupling.

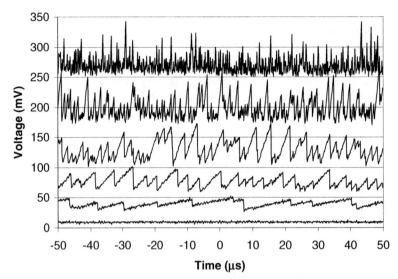

Figure 3.21   Time response of the ZTX453 noise-generator. From bottom to top the reverse DC current is 0, 0.2, 1.1, 2.95, 9.5, 31.5μA. Increasing current increases the noise bandwidth.

3.22). Eventually, at very high bias, the signal becomes quieter again. The frequency shifts can also be seen on the spectrum analyzer. At higher bias currents useful noise power is generated even at several megahertz. For measurements over the DC-100 kHz band a 10 μA current gives an essentially flat noise spectrum. I have found no good explanation for these effects but the reverse biased emitter-base junction is a good place to start. Pnp devices perform similarly.

It is also possible to generate noise-like signals using digital circuitry. National Semiconductor has developed the MM5837 and MM5437, 8-pin ICs that generate a pseudorandom bit stream (PRBS). After filtering in a 10 k/680 pF low-pass filter, the output approximates white noise. You can also build your own PRBS generator. These are usually configured as a long digital shift register with feedback via logic gates. They are designed using interesting modulo arithmetic, with each shift register stage representing one power of a long polynomial expression. The feedback taps represent the polynomial coefficients (Hickman, 1999). An example is shown in Fig. 3.23, based on a CD4006 CMOS 18-bit static shift register and a CD4070 quad exclusive OR IC. One of the gates is also used as a clock generator. For a shift register with N stages (here $N = 17$), the bit stream consists of $2^N - 1$ pulses, one pulse per input clock pulse, before repeating again. In the time domain, the stream is a sequence of 1s and 0s, usually with one string of N 1s or N 0s being forbidden, but otherwise all other possible N-bit sequences being represented equally. As $2^N - 1$ is

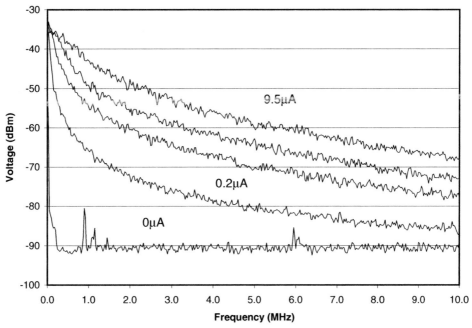

Figure 3.22 Frequency spectrum of the ZTX453 noise generator at current (bottom to top) of 0, 0.2, 1.1, 2.95, 9.5 μA.

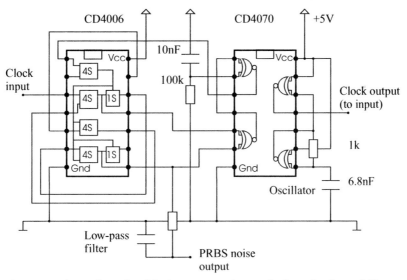

Figure 3.23 A pseudorandom bit stream generator can be formed using a shift register and a few gates for feedback.

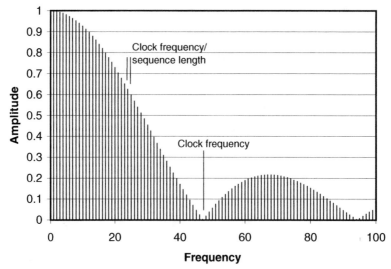

Figure 3.24 The frequency spectrum of the PRBS generator consists of a series of lines separated by 1/(sequence repeat time), modulated by the sinc(x) waveform with its first null at the clock frequency.

odd, the number of 1s and 0s differs by one. In the frequency domain we obtain a large number of discrete lines (Fig. 3.24), harmonics of $f_C/(2^N - 1)$, where f_C is the clock frequency, with a sinc(x) = sin(x)/x envelope function with its first zero at f_C. Of course the binary sequence in the time domain is nothing like Gaussian noise. To approximate that we need to low-pass filter the bit stream, typically with a break frequency of about 10 to 20 percent of the clock frequency.

Figures 3.25 and 3.26 give the measured frequency spectrum for the circuit of Fig. 3.23, a 17-stage unit, before low-pass filtering. With a wide-band view you can see the sinc(x) function, but not the unresolved lines. At high resolution the individual components become visible. It is driven here at 840 kHz, so the first spectral null is at 840 kHz, and the frequency components repeat every 6.41 Hz. A few more shift-register stages will increase the spectral line density, soon making their resolution difficult.

Last, we can of course use our photodetector as an adjustable-output noise generator. We already showed something like this in the circuitry of Fig. 3.7. Problems with this approach include the frequency limitations given by AC coupling, by the time constant of the load resistor, and photodiode parasitic capacitance. This is, after all, a receiver, and designing a wide-band noise generator should be just as difficult as a wide-band signal receiver. The circuit in Fig. 3.27 is an improvement. Here we use the same quiet DC drive of an LED to produce a high-brightness source, which illuminates two photodiodes that are unbiased but connected back to back. With exactly balanced illumination

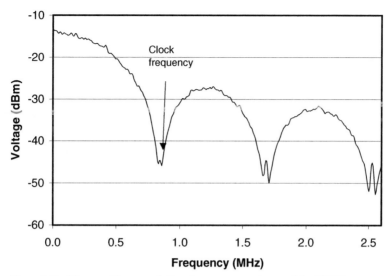

Figure 3.25 Measured low-resolution frequency spectrum of the PRBS generator of Fig. 3.23.

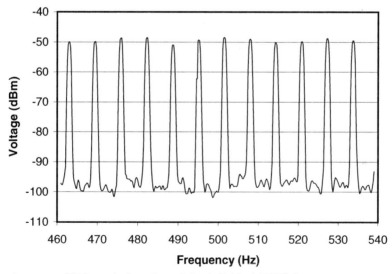

Figure 3.26 High-resolution view of the individual PRBS frequency spectrum lines, here separated by 6.41 Hz.

levels equal DC photocurrents flow through both diodes and the opamp output is zero. The shot noise signals are, however, uncorrelated and flow through the transimpedance to give a voltage output. Careful balancing is best done by mounting the LED on a micrometer stage, adjusting for zero DC output from

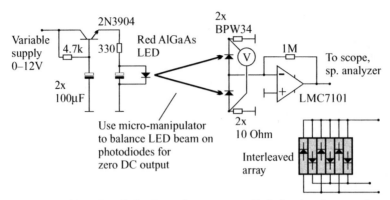

Figure 3.27 Two-photodiode shot-noise generator. By balancing the two photocurrents the opamp output is at zero volts but shows the added noise power of the two photocurrents. An illuminated interleaved isolated diode array might be a useful noise source.

the opamp. The noise power is two times that of a single diode ($\sqrt{2}$ times the noise voltage). By varying the LED brightness it was possible to vary the noise almost from the noise level of the $1\,\text{M}\Omega$ resistor (–124.9 dBm in 1 Hz) up to the shot noise current of the two photodiodes passing $200\,\mu\text{A}$ (–85.9 dBm), coupled though the $1\,\text{M}\Omega$ load. With more or brighter LEDs, or use of an optical configuration with improved coupling to the photodiodes, even higher photocurrents and noise powers should be achievable. The actual photocurrent required for calibration can be obtained from a sensitive voltage measurement across small load resistors connected in series with the free photodiode leads (Fig. 3.27). Perhaps an array of interdigitated photodiodes (inset of Fig. 3.27) would make a good, practical, single-component noise generator. It might operate with adequate balance even without the use of manual balancing. Of course if wide-band white noise operation is needed, some effort should be spent to provide the receiver with adequate bandwidth, using all the techniques discussed in Chap. 2.

3.13 Summary

With its fundamental and its avoidable contributions, the difficulties of obtaining meaningful measurements at low levels, disagreement about what kinds of noise should appear under what conditions, and the need to calculate in complicated circuits with marked frequency dependence, noise is a complex subject. It is even possible that new measurements can show new effects not seen before. Nevertheless, noise is so central to the performance of an instrument or experiment that it is essential that it be estimated at the start, calculated in detail as the design progresses, and measured in practice as soon as hardware is available. When just 3 dB more power from an LED, laser, or plasma source often

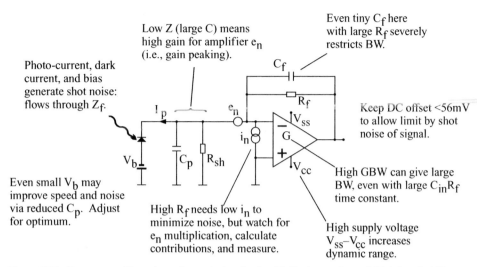

Figure 3.28 Summary of transimpedance design trade-offs for low noise and high bandwidth.

comes at such high cost, it is important that every small signal loss and drop in efficiency is recovered, if possible, and that every decibel of discrepancy in signal-to-noise is accounted for. We end this chapter with a little review of the transimpedance amplifier (Fig. 3.28). It is not the only circuit possible where high sensitivity or speed are needed, but it is generally worth consideration. Although it doesn't really improve noise over a biased detector and resistor, it does make life much easier by increasing bandwidth. We have shown that the speed of a receiver using a high-capacitance detector can be increased substantially through the use of an amplifier with a high gain-bandwidth product. However, this must be balanced against the enormous factor of amplifier noise multiplication that will occur with low detector impedance and high transimpedance. As usual, one rarely gets something for nothing, so we had better calculate and measure to understand the trade-offs.

4

Interlude: Alternative Circuits and Detection Techniques

4.1 Introduction

The majority of photoreceiver designs fall into a couple of simple configurations, usually a reverse-biased photodiode with voltage follower or transimpedance amplifier and resistive load. However, these are not the only ways to detect light and design receivers, and we should be open to alternative approaches. In this section we look at a handful of less common but occasionally very useful photodetection configurations.

4.2 Optical Feedback Systems

We have pointed out the considerable difficulty of choosing the transimpedance feedback resistor in some applications. Perhaps the high values required are not available, they are too expensive, their temperature coefficients are too large, they show excessively high parasitic capacitance, or they are physically too large for the application. This begs the question: what is a resistor? We could be pedantic and say that it is a *two-port linear network for converting a time-varying voltage into a proportional time-varying current*! There are many other ways to perform this function. Figure 4.1a shows a transimpedance configuration drawn "upside down," with its "two-port linear network" underneath. Rather than considering the output voltage as the result of the photocurrent, here we say that it is the voltage $V_o(t)$ that drives the photocurrent $I_p(t)$ through the transimpedance network and photodiode. Below the transimpedance are four circuit fragments that perform the same function.

Figure 4.1b is just the transimpedance itself. Figure 4.1c uses the voltage $V_o(t)$ to drive an LED, part of whose output generates a photocurrent $I_p(t)$. Under the above definition the resistor, LED, photodiode combination could equally be

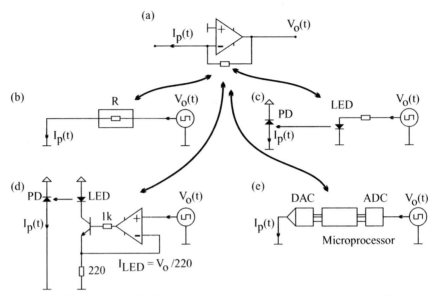

Figure 4.1 Alternative realizations of a resistor for use in transimpedance amplifiers.

called a resistor. Figure 4.1*d* attempts to correct the nonlinearity of $V_o(t)/I_p(t)$ by driving the LED with an operational current source, illuminating the photodiode as before. Figure 4.1*e* shows another "resistor" embodiment for digital fans.

The idea of synthesizing a resistor using a source/photodetector pair has been reinvented several times since about 1971 and can be applied very usefully to high-performance photodetection systems. Figure 4.2 shows a transimpedance amplifier with optical feedback. The transimpedance resistor has been replaced with an optical feedback path. This approach can significantly improve performance where receiver bandwidth and noise performance are limited by the parasitic capacitance of the feedback resistor. The capacitance and thermal noise of the feedback resistor have been removed completely. Although the feedback photodiode adds to the capacitance at the amplifier input, its effect can be small. Typically, a very low capacitance chip photodiode whose capacitance will be swamped by the signal photodiode and input FET gate capacitance will be used. There is still the dark current to contend with, and both photodiodes will show the full shot noise of the detected photocurrent. Dark current will be minimized through the use of a very small, low-leakage silicon device for the feedback, even if the signal photodiode must use ternary semiconductors for operation beyond $1\,\mu$m wavelength. In addition, the chip photodiode can often be physically much smaller than a high-value resistor and can benefit from reduced microphonic sensitivity and electrical leakage.

The next issue is dynamic range, or the ratio between the highest and the lowest detectable powers. This is important in many systems with large variations in transmitter-to-receiver loss such as optical fiber networks that have

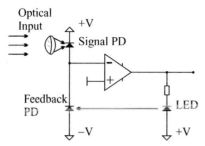

Figure 4.2 An optical feedback trans-impedance amplifier.

to be reconfigured periodically. For high sensitivity we choose high-value resistances, but unfortunately they have limited current handling ability. The product of photocurrent and resistance has to be less than the supply voltage used. With optical feedback, high photocurrents can be reduced by the LED photocurrent. This effectively compresses the dynamic range needed. In practice the optical feedback receiver offers much wider dynamic range.

Kasper et al. (1988) have described optical feedback systems operating at $1.55\,\mu$m and 1.544 Mbit/s. Feedback was provided by a $1.3\,\mu$m laser operated as an LED below threshold. Forsberg (1987) described a simple silicon-detector system with optical feedback for use at 5 MHz. In the laboratory the alternative of changing the photodiode load is equally attractive and combined with a programmable variable source power can cover a wide dynamic range (e.g., pW-mW receiver, with source electronically variable over three further orders of magnitude).

4.3   Capacitive Feedback Systems

We have seen how the problem of obtaining large value, low parasitic capacitance transimpedance resistors for the detection of very weak signal photocurrents has stimulated development of optical feedback systems. Here we will look at another option; if capacitance is the real difficulty, it is tempting to do without the resistor completely and use pure capacitive feedback. The transimpedance configuration then becomes a pure integrator.

Figure 4.3 shows such an integrator, formed by replacing the transimpedance resistor with a pure capacitance. With an input photocurrent $-I_p$ and integration capacitance C_{int}, the voltage at the output of the first opamp will rise linearly at a rate I_p/C_{int} V/s until, with this polarity, the positive supply rail is reached. The rate of rise tells us the photocurrent. Of course, it is necessary to include a switch to periodically reset the integrator. The switch can be a manual device, electrically operated relay or reed switch, or FET semiconductor switch.

There are several ways in which this circuit can be operated. As shown in Fig. 4.3, during the signal integration period the photocurrent is connected to

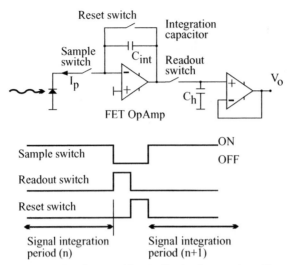

Figure 4.3 Purely capacitive feedback photoreceiver. Photocurrent is integrated on C_{int}, and then read out using logic driven switches. Adapted from Liu et al. (1993).

the amplifier's virtual earth, and its reset and readout switches are opened. After a fixed period the sample switch is opened and the readout switch is closed to transfer the voltage on C_{int} onto the holding capacitor C_h. After the voltage on C_h has been read out nondestructively via a high input impedance voltage follower, the reset switch is pulsed momentarily to discharge C_{int}. The process can then be repeated. The output then consists of a series of samples of the integrated charge performed at a fixed rate. As long as the integrator is not allowed to saturate, the sampled voltages are proportional to the integrated intensity during the integration period. The circuit is convenient in allowing the sensitivity to be adjusted over a very wide range by varying the capacitance C_{int} and the time between reset operations.

Alternatively we can operate the circuit asynchronously, resetting only when a threshold voltage is reached. Here the output signal can be the time taken to reach threshold or the frequency of resetting. The circuit becomes a photocurrent-controlled oscillator. Integrated light-controlled oscillators are available from Burr-Brown and TAOS.

The circuit in Fig. 4.3 can easily be fabricated using conventional opamps, discrete capacitors, and MOSFET switches. However, a purpose-built integrated circuit is available for this function. The Burr-Brown ACF2101 includes two complete integrators as in Fig. 4.3 contained in a 24-pin package. Each has a 100-fA bias current opamp, the three FET switches, and a 100 pF integration capacitor. By integrating two complete systems into a single unit, highly matched measurements can be performed on, for example, a signal and a reference channel.

The paper by Liu et al. (1993) describes experiments made with such a system of highly stable measurements of optical absorption for analytical chemistry. Peak-to-peak noise levels as low as $3\,\mu$AU were demonstrated (AU = Absorbance Unit; $1\,\mu$AU $= 2.3 \cdot 10^{-6}$ intensity change). Hard-wired logic circuitry was used to sequence the MOSFET switches. An alternative would be use a small microprocessor such as a PIC, or even the more convenient BASIC Stamp from Parallax Corporation. This has a built-in interpreter for a high-level (BASIC) program and can very simply be arranged to drive the three switches using the Stamp's "high", "low", and "pause" commands. For low-speed applications with minimum pulse times of several milliseconds and hence a data rate of a few measurement per second the rapid in-circuit reprogrammability of the Stamp is ideal. If higher speed is required a compiler is more suitable (such as one of the several packages provided for programming PIC microprocessors).

Although more complex than a continuous-time transimpedance configuration, the current-integration technique has a number of advantages. We have mentioned the ability to easily vary sensitivity via the integration period. With digital control and a crystal-controlled clock the integration period can be chosen to be precisely 1/50 s or 1/60 s, giving good suppression of signals at the 50/60 Hz line frequencies.

By using a capacitor instead of a transimpedance resistor, we can in principle avoid thermal noise. Purely reactive components have no thermal noise, which is only generated by the real part of the component's impedance, such as lead resistance, leakage, and inductive eddy current losses. Capacitive transimpedance amplifiers have also been widely investigated for reading out active pixel CMOS image sensors. In this way very small photocurrents can be determined with good performance. See for example Fowler et al. (2001).

4.4   Forward-Biased Photodiode Detection

All the circuits described and analysed in this book so far operate the photodiode unbiased or reverse biased. However, this is not absolutely necessary. Figure 4.4 shows the photodiode characteristic curves under various conditions of illumination, originally shown in Fig. 1.9. While normal photodetection takes place in the third quadrant, and the fourth quadrant illustrates solar cell operation, photodetection in the first quadrant is equally possible.

As previously demonstrated, every junction diode exhibits some photosensitivity, including the common LEDs. Half-duplex optical fiber communication systems have been built in which an LED was used sequentially as a source and photodetector. Figure 4.5 shows one application of a self-detecting LED as a sensor for an encoder or tachometer. The LED is operated under strong forward bias from a load resistor R_L. The emitted light illuminates a small region of a retroreflective code plate with nonreflecting stripes. A small fraction of the emitted light is reflected back into the LED chip where it creates carriers in the usual way. Although the wavelengths of peak emission and peak photodetection sensitivity for the LED do not occur at the same point, the overlap of the two

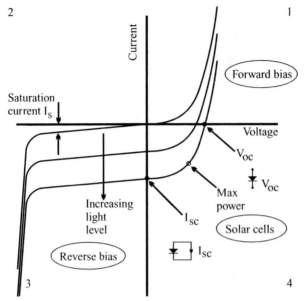

Figure 4.4 Photodiode IV characteristic. Photodetection is possible in quadrants 1, 3, 4.

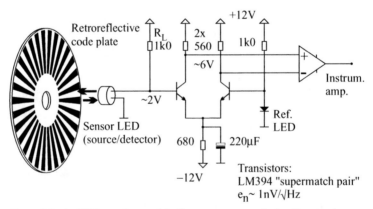

Figure 4.5 An LED can be used both as a source and simultaneously as a detector of its own light, for example to detect a rotating reflective encoder. For low-noise detection the transistor amplifier must be chosen for lowest e_n, lowest base spreading resistance r_{bb}, and high collector current.

spectra is significant and useful. The LED can therefore be used *simultaneously* as a light emitter and a photodetector.

The device shown in Fig. 4.5 was conceived for applications similar to the common gapped source/detector optocouplers and as high-resolution encoder readers embedded in the "flying" heads of magnetic storage disk systems. As the size of these devices needs to shrink, it becomes increasingly difficult to integrate both an LED and a photodetector into the same package. The LED was used as the primary sensor for track servoing to read out reflective tracks printed onto flexible disk media.

If you try this by just biasing the LED and looking at the LED voltage AC coupled to an oscilloscope input you are unlikely to see any signal. With proper receiver design, however, signals can be detected. The rules are exactly the same as those used in Chap. 3, although the requirements and design solutions are rather different from the high-resistance, high-sensitivity designs of the discussion up to now.

The forward-biased LED photodetector passes a static current of several milliamps, and we have to contend with the resulting full shot noise. The photocurrent generator acts in parallel with the forward-biased perfect diode of our photodiode model. Hence it acts as a generator with very low impedance, approximately the few ohm dynamic impedance of the diode ($25/I(mA)\,\Omega$), and a receiver must be optimized to work with this low source impedance. Similar requirements exist for low-noise detection of signals from moving-coil record player heads, moving-ribbon microphones, and certain types of photoconductors.

The dominant parameter determining noise performance here is the *voltage noise* density of the input amplifier stage. Most operational amplifiers have voltage noise spectral densities in the range 10 to $50\,\mathrm{nV}/\sqrt{\mathrm{Hz}}$, but better performance can be obtained with discrete transistors. As we saw in Chap. 3 the input-equivalent noise sources of junction transistors are the shot noise of the base current and the thermal noise of the effective base resistance. These are given by:

$$i_n^2 = 2qI_b = \frac{2qI_e}{\beta} \tag{4.1}$$

$$e_n^2 = 4\,kT\left(\frac{r_b + r_e}{2}\right) : r_e = \frac{kT}{qI_e} \tag{4.2}$$

I_b and I_e are the transistor base and emitter currents, and r_b, r_e are the base spreading resistance and emitter small-signal resistance. β is the transistor current gain. Figure 4.6 shows the variation of i_n and e_n with transistor emitter current.

For the $\approx 1\,\Omega$ source resistance of the forward-biased LED, the current shot noise is negligible and thermal noise is dominant. The design clearly needs the input transistor to be run at high current. When this is optimized, signal-to-noise is limited by its base-spreading resistance r_{bb}. This can be minimized by connecting several transistors in parallel and/or by choosing devices specially processed for low r_{bb}. Large die power transistors are sometimes used for this purpose, but specialist devices are better. Devices such as the 2SD786 and 2SB737 transistors are available with r_{bb} of $4\,\Omega$ and $2\,\Omega$ respectively, which leads to $e_n < 1\,\mathrm{nV}/\sqrt{\mathrm{Hz}}$. Figure 4.5 shows a high-gain common emitter amplifier designed around the LM394 bipolar supermatch pair. A balanced configuration using a second LED reduces the effects of temperature drifts, but the circuit is

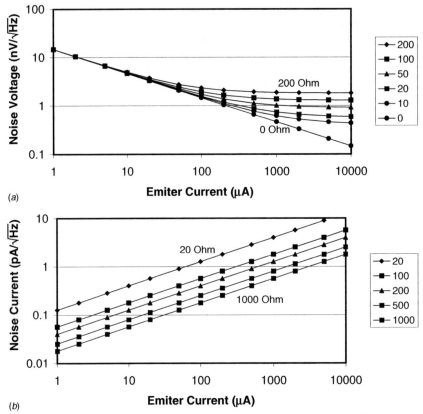

Figure 4.6 (a) Bipolar transistor voltage noise generator for various values of r_{bb}. (b) Bipolar transistor current noise generator for various values of emitter resistance r_e.

still restricted to AC coupling. Figure 4.7 shows a time-trace of a 660 nm (red) LED operated in close proximity to a spinning disk of retroreflective adhesive tape masked by opaque black stripes. Eight black stripes of different widths are visible. Most of the visible noise is due to true variations in surface reflectivity. The signal-to-noise for detection is more than 35 dB in 50 kHz bandwidth.

4.5 Wavelength Shifted Detection

Some regions of the wavelength spectrum are difficult to detect. For instance, above about 700 nm wavelength the eye becomes less and less sensitive, which makes working with the near-infrared light (IR) of current optical communication systems difficult. The well-known IR viewing cards are used to make visible the energy in this region, from about 800 to 1600 nm. They function by storing visible energy from room lights and sunlight in a layer of phosphor material, which is then stimulated by the IR to be released in the visible spec-

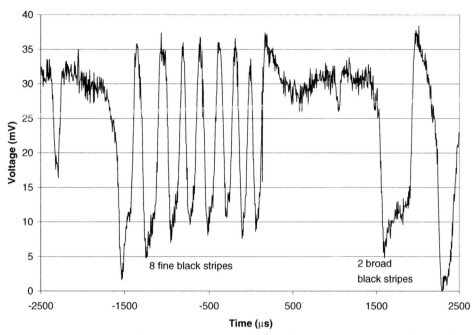

Figure 4.7 Self-detecting LED output for eight 1-mm-wide black stripes on a retroreflecting disk.

trum. Emission wavelengths are usually in the 610 to 640 nm range. In principle they could be coupled with a silicon detector to give improved sensitivity beyond 1 μm wavelength, but in practice it is far better to use a germanium or InGaAs photodiode.

Below 350 nm most silicon detectors also show poor and decreasing responsivity, due both to use of a borosilicate glass or plastic window and to the intrinsic detection process. Especially at these ultraviolet wavelengths, therefore, fluorescent wavelength conversion is an attractive option. The goal is to find a material with high absorption at the "difficult" wavelength, which reemits in a longer-wavelength region where the photodiode sensitivity is superior (Fig. 4.8*a* and *b*). If the emission process is efficient, there can be a net gain in detection sensitivity. A number of fluorescent glasses, polymers, and dyes is available for this. For example, Sumita Optical Glass manufactures a range of fluorescent glasses that can be mounted directly on the front of the photodetector as a wavelength converter. Some can also be processed into planar waveguide form with edge-mounted detectors, allowing large areas for detection of diffuse or widely dispersed light (Fig. 4.8*c*). Large-core (mm^2) optical fibers can be formed from some fluorescent polymers. The best materials for this use show high absorption at the detected wavelength and low absorption at the shifted wavelength, so that self-absorption of the visible light is minimized. With the great interest in and potential markets for efficient white-light LEDs, it is likely that big improvements will be seen in phosphor materials for use at deep blue/near-UV

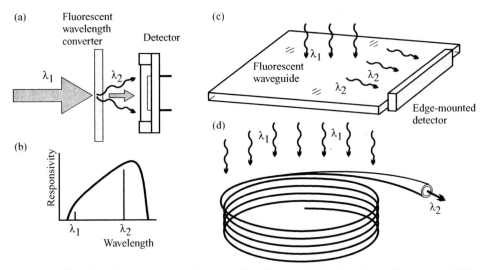

Figure 4.8 Wavelength conversion can improve detection sensitivity in regions of poor photodiode responsivity. Fluorescent and scintillating materials can be formed into face plates, planar waveguides, and fibers.

excitation wavelengths. Doped glasses researched for optical fiber amplifier use have similar properties. Both of these may find application with detectors. A shard of phosphor-coated glass from a broken fluorescent tube can be used for efficient detection of 254 nm UV light. When detecting rapidly modulated light remember that phosphor wavelength conversion can be slow.

When the target wavelength becomes shorter still, fluorescence makes way to scintillation. Thallium-doped cesium iodide is the most common scintillator material, which generates a smooth 200 nm wide visible output peak centered on 680 nm, when illuminated with 100 keV x rays. Synthetic ceramic materials are also available, emitting a series of lines at about 520 nm, 68 nm, and 780 nm, with similar sensitivity but reduced afterglow. Hamamatsu manufactures a range of silicon photodiodes with these materials already bonded as faceplates. The primary application is in x-ray tomography.

Wavelength conversion can also be a problem for photodetection. When working with low-pressure mercury vapor lamps, which emit at 254 nm, or the UV lasers used for fiber Bragg grating manufacture, it is difficult to avoid fluorescent wavelength conversion. At these wavelengths, almost everything seems to fluoresce, including many glasses, polymers, dyes, aromatic chemicals, biological materials, and fabrics. Even ordinary white bond paper makes an effective viewing card for the UV due to additives. Fabric "brighteners" contained in clothes washing powders emit intensely when excited by a violet LED or UV lamp. If it is intended to make quantitative absorption measurements in the UV, this extra light can lead to large errors and peculiar results. When pulsed light sources are used and detected, the long fluorescent time constants of efficient materials will also distort pulse shapes. To quantify the UV signal free from these effects, it may be necessary to use a narrow-band UV interference filter in front of the detector.

Alternatively, you might need to suppress the excitation wavelength and detect only the fluorescence. This is the requirement for fluorescent detection used in chemical analysis. It is the basis of many of the most sensitive assays possible. The big benefit, as discussed in more detail in Chap. 10, is that fluorescent detection provides a *dark-field* measurement. Unlike the case of trace-level detection of chemicals by transmission spectroscopy, where almost all the light passes straight through the sample and we must detect tiny reductions in transmitted intensity, in fluorescence detection when no sample is present, we should have no detected light. This allows the detection gain to be greatly increased to improve the limit of detection (LOD). The key to good performance in fluorescence measurement is perfect separation of the excitation and converted wavelengths (Fig. 4.9). Unless a very well-defined wavelength is used for excitation, spurious longer emission wavelengths may be confused with the fluorescent signal. Similarly, at the detector the excitation wavelengths must be very well suppressed. This usually demands two sets of filters, one to clean up the source and one to limit the detection bandwidth. Multilayer interference and holographic filters are commonly used, although liquid and absorbing glass filters can be useful in some cases, as interference filters have additional passbands at longer wavelengths. In addition to requiring excellent filtering, because the fluorescent processes emit isotropically, it is also desirable to collect

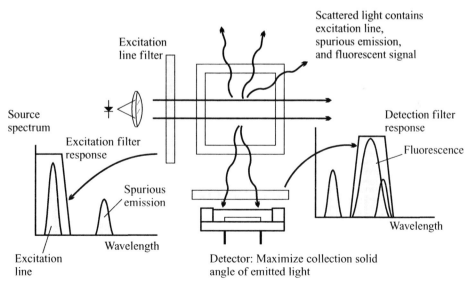

Figure 4.9 Sensitive detection of fluorescence requires well-defined wavelengths of excitation and detection. This generally requires both source filters (to remove fluorescence wavelengths) and detection filters (to remove the source light wavelengths).

light over as large a solid angle as possible. This either requires large detectors or very compact systems. As usual, this leads to difficult design trade-offs. Nevertheless, fluorescence can deliver superb performance. In liquid chromatographic detection a mass LOD of a few femtogram (fg) of material is possible, about 100 times better than using absorption detection. See for example Skoog et al. (1998) for comparisons.

4.6 Parametric Detection

When we want to detect tiny changes in optical transmission in materials that are largely transparent, we have extra difficulties. This is a small change on a large background, and we must first contend with the shot noise of the full-scale transmitted signal. Second, as discussed in Chap. 10, the best sensitivity is often limited not by measurement noise but by variations in source intensity. This problem can be ameliorated by using techniques that measure not the transmitted light but some light-influenced property. The most widely studied methods belong to a family of photothermal techniques.

One approach is to detect the light actually absorbed by the sample via the heat generated in absorption. We could use any type of thermometer, but for solid samples a more sensitive technique might be to measure thermal expansion using a sensitive interferometer (Fig. 4.10a). If the "absorbed beam" is slowly scanned in wavelength through an absorption feature, the interferometer will in principle read out the absorption through changes in temperature and expansion. The absorbed beam here performs the job of the light source of a spectrometer. However, in doing this we have suppressed the large signal when absorption is lowest. With the expansion readout, no absorption means zero signal. We have also separated the excitation and readout optics. To get high sensitivity we can increase the intensity of the probe beam to improve the shot-noise limited S/N, and operate in a wavelength region with low-noise photodetectors. For liquid samples we can additionally make use of a hydraulic gain. Figure 4.10b shows a system in which the sample expansion is detected via movement of a liquid meniscus, a liquid version of the famous Golay detector. A near-infrared-sourced single-mode fiber interferometer was used to sensitively detect the liquid expansion caused by UV and blue light absorption (Hodgkinson, 1998). As long as absorption and heat generation in the transparent windows and at the cell wall are minimized, this technique also allows us to separate absorption from scattering. In a conventional spectrometric transmission measurement these two are mixed and cannot be directly separated.

Temperature changes can also be transduced into refractive index changes, which can be read out to high resolution using beam deflection techniques. Figure 4.10c shows detection of the refracted beam caused by a hot spot in the sample cell using a split photodiode. Without absorption the probe beam passes straight through to impinge symmetrically on the photodiodes. With absorption the beam is deflected through a small angle to unbalance the two photocur-

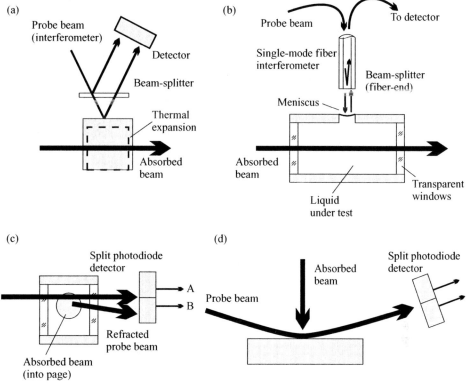

Figure 4.10 Optical absorption detection by (*a*) thermal expansion of a solid, (*b*) thermal expansion of a liquid, (*c*) refractive index changes in a liquid, and (*d*) in a gas (mirage effect).

rents. Note that this is not a dark-field measurement, where zero absorption gives zero photocurrent, as a high intensity is detected under all circumstances. Commonly the two photocurrents *A*, *B* are detected separately in transimpedance amplifiers, and the output signal is the photocurrent difference normalized to the total intensity:

$$S = \frac{A - B}{A + B} \tag{4.3}$$

With a high-power readout beam the shot-noise limited S/N can be high and hence high sensitivity obtained. The two shot-noise signals are uncorrelated and so add as sum-of-squares. Figure 4.10*d* shows another approach called *mirage-effect* detection. Here the scanned beam is absorbed near to the surface of a solid or liquid sample, and some of the probe beam is refracted in the temperature gradient formed in the air close by. Again, this can very sensitively detect temperature changes at the surface of a solid sample and hence very weak absorptions such as in laser mirrors.

The methods of Fig. 4.10 are primarily static techniques, because the thermal time constant of the sample heating process is rather long. However, if the absorbed beam is modulated, it is possible to detect the modulation as sound. The techniques are called photoacoustic detection photometry. The sound field is still generated by expansion, so it is advantageous to use solvents with a high expansion coefficient. In water this is maximized at 4°C. For gas detection conventional electret or condenser microphones can be used. For liquids and solids custom-built piezoelectric transducers are more common. The majority of published work uses high-power pulsed lasers to excite the absorption, although it is possible to obtain similar performance with continuously modulated sources, including LEDs (see Hodgkinson, 1998). This broadens the range of wavelengths at which absorption can be measured without going to the complexity and expense of tunable lasers. Using a microphone detector the technique is dark-field, with zero signal at zero absorption. In principle, all these parametric techniques can deliver higher detection signal-to-noise than can direct detection. In practice, this is far from easy due to spurious absorption in optical windows, so that equivalent performance can usually be obtained using high stability and low-noise transmission measurement or the measurand modulation techniques of Chap. 10.

System Noise and Synchronous Detection

5.1 Introduction

In Chap. 3 the fundamental noise contributions to our signal, shot noise of currents and thermal noise of resistors, were discussed. These noise sources in general have no particular spectral character, delivering the same power per unit bandwidth up to a very high frequency. However, the typical noise power spectrum seen in a real photoreceiver in a real environment is far from being as smooth as the quantum analysis leads us to expect. First, we have seen how the frequency-dependent networks of electronic amplifiers and other components connected to the photodiode can significantly modify the noise density as a function of frequency. Further, the majority of semiconductor devices and even many *processes* exhibit a $1/f$ character, which greatly increases noise density at low frequencies, typically below about 500 Hz. The voltage and current noise spectral densities of both discrete transistors and opamps show this $1/f$ character, with a corner frequency in this region.

Many man-made sources of electrical and optical interference are present at the output of an optical receiver. At 50/60 Hz, 100/120 Hz, and 150/180 Hz there are often very strong electrical interference signals from line-voltage wiring and power supply transformers. They can usually be greatly suppressed, if not totally eliminated, by good electrostatic and/or ferromagnetic screening of receivers, through diligent bypassing of power supply and signal leads entering the receiver and through the avoidance of ground loops. In addition, many other natural interfering light sources find their way to our detectors that vary in intensity at a low frequency. These include sunlight (1/24 hour frequency), sunlight modulated by clouds ($\approx$1/minutes frequency) and flashing television and computer screens (25 to 120 Hz frame rate, 15 to 150 kHz line rate).

All this suggests that measurements should be made, if possible, far away from the low-frequency region 50 to 500 Hz. Electrical interference can be fairly quiet in the 3 to 50 Hz region, but $1/f$ noise is large and the low frequencies make measurements done there rather slow. Shifting the measurement of optical power to a frequency above this range is usually preferable, and is the job of modulation.

5.2   Modulation and Synchronous Detection

Instead of working DC with a photoreceiver connected to a voltmeter, in a modulated system the light source must be modulated, for example in intensity, and only the detected AC signal measured. Where the source is a simple one, such as a light-emitting diode (LED) or laser diode, modulation is usually straightforward to arrange. Simply connecting an LED via a current-limiting resistor to an audio-frequency, square-wave, or pulse generator will effectively modulate the LED's light output. If the generator output is bipolar, it is safer to connect a silicon diode in inverse parallel with the LED to avoid exceeding the latter's reverse voltage limitation. If sinusoidal modulation is desired, the LED should be biased with a DC current of about half its maximum current and modulated about this current. Chapter 6 gives a few guidelines on these practicalities.

The detected, buffered signal may be viewed and measured visually using an oscilloscope, or after AC coupling using the AC voltage ranges of a standard voltmeter. Most analog and digital voltmeters will still measure an AC signal at a frequency up to a few hundred hertz or so, giving an average or rms reading depending on the circuitry and calibration. This may be an adequate approach if the signal is strong and the signal-to-noise (S/N) high. However, the AC voltmeter, with its ill-defined and probably wide detection bandwidth, does not make best use of the properties of the modulated light source. For better performance and flexibility in choice of detection bandwidth one should use synchronous detection.

"Synchronous detection" is a fancy name for changing the source intensity and then looking for the change in detected output you expect to see; this is a very powerful principle, and not just in electronics. Electronically, this is performed as in Fig. 5.1. The figure shows an LED driven by a current generator

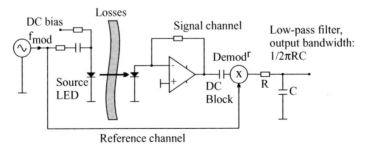

Figure 5.1 Synchronous detection requires a reference signal that is phase-synchronous with the signal light and that is multiplied by the received signal. The combination of modulator and low-pass filter forms a bandpass filter centered on f_{mod}.

to give a sinusoidally varying light output, with frequency f_{mod}. The light, perhaps after suffering large loss by absorption, is detected by an optical receiver, in this case in transimpedance configuration. The detected signal, which contains both the weak modulation due to the LED's light as well as DC due to amplifier offsets and perhaps static ambient lighting, other disturbing AC signals, and of course noise, is applied to one input of an analog multiplier. Although not essential, the signal is shown here AC-coupled, to remove at least the DC offset and the lowest frequency signals. The other multiplier input comes from the same sine-wave generator used to power the source. After the multiplier the product is filtered in an RC low-pass filter.

At a frequency well above its $1/2\pi RC$ characteristic frequency, the low-pass filter looks like an integrator. Hence this circuit fragment forms the product of our signal with a sine wave and integrates it. Mathematically, the action of this synchronous detector is similar to that of computing the sine Fourier transform. That technique is used to determine the amplitude of a particular frequency (f) component of an input signal. If the input signal which varies with time t is $S(t)$, we compute the integral:

$$\int S(t) \sin(2\pi ft)\, dt \tag{5.1}$$

In similar fashion the structure of Fig. 5.1 determines the amplitude of the sinusoidal component of our input signal. If the signal is still approximately sinusoidal, and in phase with the reference, it determines the amplitude of the signal itself.

In the frequency domain (Fig. 5.2), the synchronous detector functions as an electrical filter, centered on f_{mod}. The passband width of this filter is determined by the RC product of the low-pass filter and is of the order of $\pm 1/2\pi RC$ Hz. It is as though the one-sided, positive-frequency low-pass characteristic of the RC network, with a cutoff frequency of $1/2\pi RC$, has been mirrored about the zero-frequency axis, and frequency-translated by multiplication by f_{mod} to become a two-sided bandpass distribution centered on f_{mod}.

The advantage of the modulation/synchronous detection process is clear from Fig. 5.2. We have produced an electronic filter, centered precisely on the signal we expect to find, with a bandwidth that can easily be adjusted simply through the choice of two components (R and C). The filter lets through the desired signal, but also the noise and interfering signals contained in the bandwidth $f_{mod} \pm 1/2\pi RC$. The act of translating to higher frequency should avoid some of the worst noise and interference. It is straightforward to reduce the filter bandwidth by choosing a large value for the RC time constant. A $1\,M\Omega$ resistor and a $1\,\mu F$ capacitor give a one second time constant, or a bandwidth of $\pm 0.16\,Hz$. This bandwidth will be obtained whether f_{mod} is at $1\,kHz$, $10\,kHz$, or $1\,MHz$. Achieving adequate stability of a 0.16-Hz-wide filter at $10\,kHz$ by conventional active, or passive LCR filters would be a daunting task. The synchronous filter is relatively straightforward to set up for an arbitrarily narrow passband, is as stable as the reference oscillator, and even automatically tracks a modulation

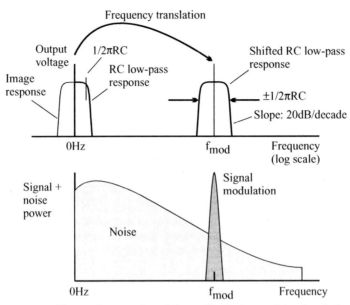

Figure 5.2 The synchronous demodulator effectively translates the single-sided low-pass filter characteristic out to a symmetrical response centered on f_{mod}.

signal drifting slowly in frequency. "Slowly" here means that the detected signal has time to settle before the frequency has moved out of the passband.

5.3 Square-Wave Demodulators and Importance of Phase

Demodulation can also be performed by multiplying the input signal with a synchronous reference clock that is a binary signal (values ±1) instead of a sine wave. This has many practical advantages, not the least of which is that it makes drawing and understanding the waveforms in the time domain much easier. We will use this representation in the discussion of detection phase. We have assumed up to now that the reference and noisy input signal are aligned with a phase difference of zero. Under these circumstances the integrated product takes on its largest positive value. This situation is shown for a binary demodulator in Fig. 5.3a. The aligned sinusoidal input is converted by multiplication into a rectified sine wave whose average value is large and positive. If the relative phase ψ between signal and reference is changed by π, the filtered, averaged output becomes large and negative (Fig. 5.3b). Figure 5.3c shows the intermediate phase ($\psi = \pi/2$), where it is clear that positive and negative areas in the product waveform are equal and the integral of the product approaches zero. Signal and reference are then said to be "in quadrature." Between these phases, the output varies sinusoidally with ψ, with the filtered output having a

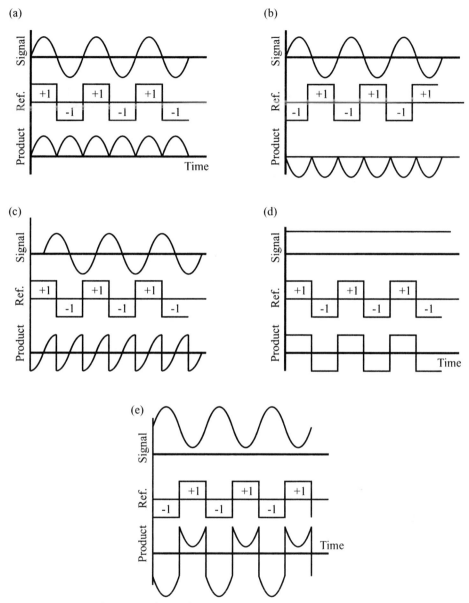

Figure 5.3 Demodulator waveforms for various relative alignment phases. (*a*) in phase, (*b*) antiphase, (*c*) in "quadrature," (*d*) zero response to a DC signal, (*e*) DC-response removed, AC detected.

magnitude of $S \cos(\psi)$. Hence the single-channel synchronous detector can be used either to determine magnitude S (if the phase is known) or phase ψ (if the magnitude is known), but not both simultaneously.

Figure 5.3d shows the response of the synchronous detector to a constant level input. The signal is chopped to give a bipolar amplitude signal whose average value is zero. This zero DC response simply corresponds to the zero of response stretching from 0 Hz to the start of the passband around f_{mod} in the frequency domain of Fig. 5.2. In Fig. 5.3e a synchronous signal is shown added to the DC level. Once again, the DC component has been removed by the chopping process, with only a finite negative average from the synchronous AC component.

Therefore it is very important that the phase difference between reference and signal has the value we need. With low-noise signals this can be achieved by adjusting the phase for a maximum postfilter signal. However, close to the maximum, the output varies only slowly with phase, like the cosine function around zero angle. Hence it is usually better to adjust the phase for a *minimum* filtered output signal, which can be much more precise, afterward shifting the phase accurately by 90°.

A common feature of synchronous measurement systems is the provision of two or more separate detection channels. If we use two multipliers driven by 90° phase-shifted signals, then the two demodulated outputs will vary as the sine and cosine of the phase angles (Fig. 5.4). Each measurement channel detects the projection of the rotating phasor R onto either the C or S axis. The magnitude of the phasor can be calculated from $R = (S^2 + C^2)^{1/2}$. Three-phase and higher-order systems can also be used and offer some advantages in terms of the symmetry of the two resolved channels.

5.4   What Frequency Should We Use?

It is clear that the modulation frequency should if possible be above the 500 Hz region of so much man-made and natural interference, but the choice is not arbitrary, as noise spectra are not flat even above 500 Hz. One of the most annoying and ubiquitous sources of optical interference is the fluorescent lighting used in most laboratory environments. These sources are more serious than might at first be thought. Although the lights are driven by a more or less sinusoidal voltage source at 50/60 Hz, the discharge process leads to a light output that is a distorted, rectified sine wave. The frequency spectrum of this waveform exhibits strong components at harmonics of the 50/60 Hz drive, which can often be seen in detected light out to several kilohertz. Figure 5.5 shows a spectral analysis of the output of a 10-kHz bandwidth transimpedance receiver with interfering fluorescent room light. This was a system designed to measure low levels of scattered light in an open environment. Despite optical filtration with a 25-nm bandwidth interference filter to remove other nonsignal wavelengths and an angular acceptance at the detector of only 5°, significant harmonics of the 100 Hz rectified line frequency are visible well beyond 2.5 kHz.

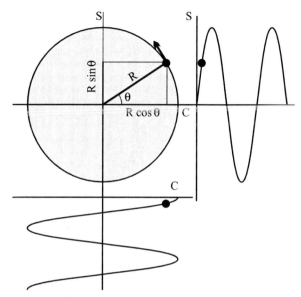

Figure 5.4 Synchronous demodulation selects one component (sine or cosine) of the rotating phasor. Sine and cosine detectors together can be used to construct the phasor amplitude R.

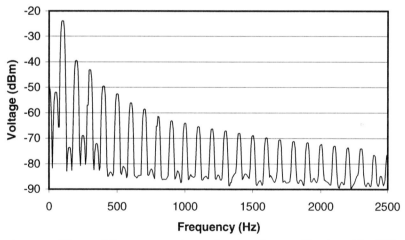

Figure 5.5 Measured frequency spectrum of a photodetector illuminated by 50 Hz fluorescent lighting, showing harmonics of 100 Hz out to 2.5 kHz.

In open optical systems it is often difficult to reduce such interfering harmonics to the receiver noise level (if it has been designed properly). When we evaluate the S/N of the measurement system, these disturbing interference signals can be just as bad as a higher noise level. In this case it is important to

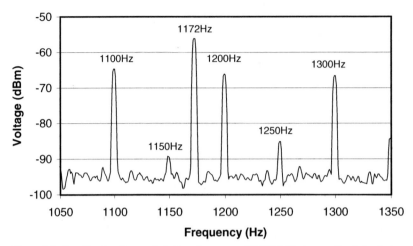

Figure 5.6 Positioning the modulation frequency (1172 Hz) between the interfering harmonics can greatly improve S/N measurement. Hitting a harmonic would degrade S/N by about 25 dB.

modulate at the "right" frequency. This might mean accurately between two harmonics of 50/60 Hz. Figure 5.6 shows an expanded spectral analysis covering 1050 to 1350 Hz. The 1100 Hz, 1200 Hz, 1300 Hz room-light harmonics are evident, as well as vestiges of 50 Hz harmonics. The modulated source signal has been deliberately placed between two harmonics, in this case at a frequency of 1172 Hz. If the modulation had been chosen instead at 1200 Hz, on top of a strong harmonic, the measurement S/N would have been degraded by at least 25 dB, perhaps the difference between a reasonable measurement and a failed experiment. Not only is the S/N greatly reduced by measuring on a disturbing harmonic, but if the reference clock drifts in frequency, the interference magnitude will change with time. This variable behavior can make diagnosis of intermittent poor S/N problems very difficult.

The only sound way to choose the frequency is to first investigate the frequency spectrum of the optical receiver's output. This can be done using a conventional electronic spectrum analyzer, by using the Fourier transform algorithms built into many modern oscilloscopes, or by capturing a time trace at high speed and performing the spectral analysis off line in a computer. Careful measurement throughout, for example of the audio frequency band, will often point out many areas that are best avoided.

5.5   Building the Synchronous Detector

At low frequencies (<1 MHz), a wide variety of analog operational circuitry can be used to construct the four-quadrant multipliers necessary to multiply two sine waves. It is easiest to use ICs such as the Analog Devices AD534/634 or Burr-Brown MPY100. However, they are complex and typically provide only about 1 percent accuracy. For this reason, many applications avoid sinusoidal demodulation, using instead the bipolar (square-wave) demodulation described above. This is simpler to perform to high precision and high speed and is therefore much cheaper to integrate.

Figure 5.7 shows one method to make a simple binary multiplier. An operational amplifier with symmetrical power supplies is combined with an FET switch connecting the inverting input to ground. If the FET channel resistance is designated R_4, the gain of this circuit fragment is $(R_1R_4 - R_2R_3)/(R_1(R_2 + R_4))$. By switching the gate voltage, its channel resistance can be switched between low and high values. If the FET resistance is $R_4 = 0\,\Omega$, the circuit gain becomes -1. With the FET switched off to the nonconducting state (infinite resistance), the gain is $+1$. Hence a square-wave drive (typically with CMOS logic levels) from the source reference generator, applied to the FET, will multiply the input voltage by ±1 at f_{mod}.

The choice of FET depends on the details of the rest of the circuitry. If the input signal is always positive (such as the output of a DC-coupled transimpedance amplifier with anode connected to ground, then an n-channel enhancement mode MOSFET can be used. These are the most common types of small MOSFET. For example, a VN2222LL MOSFET with a 0 to 5 V gate drive voltage obtained from 5 V CMOS logic ICs will perform well. Its on-state channel resistance is less than $7.5\,\Omega$, making it an almost ideal switch. As its maximum gate turn on voltage is 2.5 V, simple drive with transistor logic is possible.

If, however, the input signal varies about zero volts in a system with symmetrical power rails, the VN2222LL should not be used. This is the case with an AC coupled input. With +5 V on the gate the FET will switch on correctly to the low-resistance state. However, in the off state with $V_{gs} = 0$, negative drain voltages will start to turn on the FET, leading to incorrect operation. With enhancement mode devices it is necessary to ensure that the off-state FET does not become turned on by the signal itself.

An n-channel depletion-mode device such as the 2N4118A can also be used. With $V_{gs} = 0$ the device is on for either polarity of drain signal. $V_{gs} = -5\,\text{V}$ will

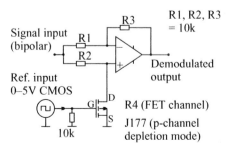

Figure 5.7 A simple binary modulator can be formed from an opamp and an FET switch. Overall gain is switched to ±1.

switch off the FET. As long as the signal is such that the drain voltage is more positive than about $-3\,V$, the switching operation will be correct. Unfortunately, most logic circuitry is traditionally arranged for positive signals.

A convenient practical configuration uses a p-channel depletion-mode MOSFET such as a J177. As this is a "normally-on" device, in which channel current flows when $V_{gs} = 0$, the on state can handle bipolar signals. The off state is obtained by putting V_{gs} a few volts positive. It needs to be at least 2.25 V more positive than any voltage along the channel. This is convenient as the gate may then be driven by 74 HC or 4000 series CMOS logic to the positive rail (+5 V or up to +15 V, respectively). The signal can then vary from the negative rail up to within a couple of volts from the positive rail without difficulty. It is best to drive this circuit from the low impedance output of an opamp.

Figure 5.8 shows an alternative approach using operational amplifiers to deliver two signals representing $\pm V_{in}$, with a MOSFET analog switch (e.g., DG419, MAX319) used to choose between the two options. The switch is driven by a digital signal at f_{mod}. The MOSFET's internal analog switches have the same problems of voltage range seen above, but built-in level shifting and more complex configurations are used to ease their application. Nevertheless, it is necessary to provide positive and negative supply rails which limit the maximum voltage excursion.

All these systems using analog switches offer some challenges in the choice of FETs, given the constraints of speed, offset voltages, gate-drive, channel resistances, resistor matching, etc. An easier solution is to use dedicated switched-input operational amplifiers such as the Burr-Brown OPA675 and Analog Devices AD630 (Fig. 5.9). This latter integrated circuit contains a pair of operational amplifiers whose outputs can be selected under electronic control. Precision matched on-chip resistors allow the IC to be wired as a precise $\pm 1\times$ or $\pm 2\times$ amplifier. It is easy to use as the heart of a very high-performance synchronous detector. In the circuit shown the clock is a CMOS square-wave AC coupled and reduced in amplitude to drive the reference input.

To guarantee an accurate 50:50 mark-to-space ratio clock, use the output of

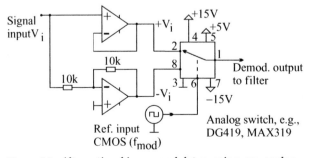

Figure 5.8 Alternative binary modulator using an analog switch to select between positive- and negative-gain signals.

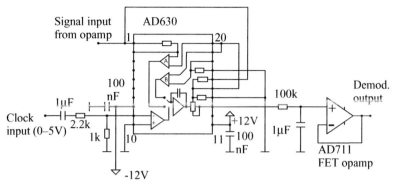

Figure 5.9 The AD630 IC can be the basis of a high-performance synchronous detection system.

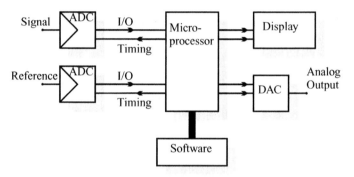

Figure 5.10 Synchronous detection, as well as more complex signal recovery algorithms, can be built using digitization followed by digital signal processing hardware and software.

a divide-by-2 circuit, such as half an HC7474 dual D-type to supply the clock. The comparator input is fairly forgiving on clock waveform. The switched output must be filtered, for example in the FET opamp RC filter as shown.

These opamp circuits are usable throughout the audio band and up to about 1 MHz. In the radiofrequency domain, operational techniques make way for simpler electronic ICs such as the MC1496 long-tailed-pair modulator. At even higher frequencies diode-ring or double-balanced mixers based on transformers and semiconductor diodes can be used. These can perform up to gigahertz frequencies. Beyond that electrooptic and optooptic switches could in principle perform synchronous detection.

At the other end of the frequency spectrum a fundamentally different approach has appeared that is becoming more widely practiced. This is the use of digital computational circuitry (Fig. 5.10). It involves digitizing the signal and reference channel voltages as soon as possible after the front-end receiver, before any analog processing except perhaps for noise reduction or antialias filtering. Once signal and reference are converted in analog-to-digital converters (ADC), all multiplications and integrations in the Fourier analysis can be carried out free from errors using digital hardware and software techniques. By taking this approach, great flexibility and richness of functions can be provided. It is, for instance, hardly more complex to compute the power in many harmonics of the signal, each in a different bandwidth, or even the full Fourier spectrum of the signal. With the relentless attack of digital solutions on areas once considered solely analog, the increasing use of digital signal processing in synchronous detection must be expected to continue. Low-frequency signals and those recorded for off-line analysis have long been dealt with this way. Even the humble radio receiver is being reduced to an aerial, ADC, and digital processing chip.

5.6 Walsh Demodulators

The ease of construction of digital ($\pm$1) multipliers hides a disadvantage of increased noise. This is easily seen from the frequency-domain representations of signal and reference. The passbands in the frequency domain formed by modulation and synchronous demodulation are obtained by convolving the frequency-domain representation of the reference signal with the double-sided passband obtained from the postdemodulation low-pass filter. The Fourier analysis of a sine-wave reference is just a single amplitude, so multiplication of the input signal by a sine wave at f_{mod} delivers a filter with a single passband, width $\pm 1/2\pi RC$ centered on f_{mod}. The Fourier decomposition of a bipolar square-wave reference clock is a set of odd harmonics of the fundamental frequency. This produces passbands at f_{mod}, $3f_{\mathrm{mod}}$, $5f_{\mathrm{mod}}$, etc., with relative amplitudes of 1, 1/3, 1/5, etc. (Fig. 5.11). Hence square-wave references open up not just a single passband, but an infinite series of passbands. If the input has signal, noise, or interference power at these harmonic frequencies, they will be detected and contribute to the overall output. It is possible to roll off the receiver to give a reduced power at three times the modulation frequency, but the suppression will not be complete. When a single RC filter only achieves an ultimate cutoff slope of –6 dB/octave (–20 dB/decade), it is difficult to get much suppression at $3f_{mod}$. Hence these harmonic responses can be a problem and must be kept in mind if strange interference effects are seen.

A partial solution to the harmonic responses can be obtained by approximating the sine function needed as reference waveform for the synchronous detector. If we can choose a small number of binary waveforms to approximate the sine wave, these can then be applied to weighted binary multipliers and summed together for an overall response similar to that of a sine wave reference. Just as Fourier synthesis can be used to approximate an arbitrary function using a small number of sine/cosine terms, we will use Walsh synthesis to approximate the sine wave with a small number of Walsh functions.

Walsh functions are two-parameter binary ($\pm$1) functions that form an orthogonal series. They can be used just like the sine and cosine series of Fourier analy-

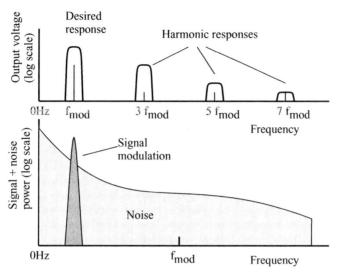

Figure 5.11 Synchronous detection using a square-wave modulator results in harmonic responses at odd multiples of the modulation frequency.

sis and synthesis to construct approximations to other functions. As the Walsh functions are inherently digital, they can be very efficient at approximating functions containing steps. (Fourier series are generally poor at this job, generating large errors at such steps: the Gibbs phenomenon.) However, they also do a pretty good job with the continuous sine functions.

The 16 Walsh functions of order four are shown in Fig. 5.12. They include both regular square waves of different periods and some nonperiodic waveforms. To make an efficient procedure, we first perform a Walsh function analysis of a sine wave. Once the coefficients of each of the Walsh harmonics are available, we will discard all but the three with the greatest amplitudes. Transforming back again gives the approximated sine wave.

Figure 5.13 shows the Walsh transform of a ±1 V sine wave, that is, the amplitude coefficients of each Walsh function term. This was performed in Mathcad, which provides a Walsh transform algorithm. The zero[th] coefficient of this analysis is just the average value of the input sine wave, and therefore is itself zero for this zero-mean sine wave. Also, all the even coefficients are zero. Further, many of the remaining coefficients are very small and can be neglected without imposing great errors. Here we take just the 1st, 3rd, and 7th coefficients, with amplitudes respectively of 0.628 –0.125 and –0.260. Adding these together we obtain a respectable, stepped approximation to the original sine wave. Figure 5.14 shows two cycles of the original sine wave, the three Walsh functions used in the approximation, and the approximated sine wave made up from the sum of the functions.

To make a Walsh function demodulator, we need only form the products of

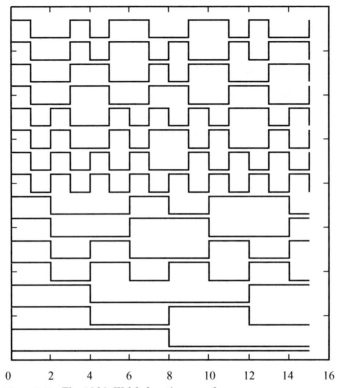

Figure 5.12 The 16-bit Walsh function waveforms.

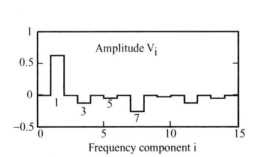

Figure 5.13 Walsh frequency spectrum of a sine wave.

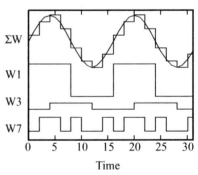

Figure 5.14 Original sine wave and the three Walsh functions used in its approximation. The stepped response is just the sum of the 1st, 3rd, and 7th individual Walsh functions with optimum amplitudes.

V_{in} with each of the three chosen Walsh functions. These multiplications are as simple as described above for a simple square-wave modulator and can be performed using the same hardware of Figs. 5.7, 5.8, 5.9, etc. To accommodate the different coefficient amplitudes, a final analog weighted-gain stage is added (Fig. 5.15). The summed result at the output of the opamp is equivalent to a multiplication of the input signal with our approximate sine wave but without the difficulty of forming a true analog multiplication.

We can see what effect this will have on the harmonic transmission bands of a synchronous detector formed in this way by calculating the Fourier transforms of the original sine wave, of the approximated sine wave, and of a square wave of the same frequency (Fig. 5.16). The third and fifth harmonic responses are seen to have been greatly suppressed (with more exact spectrum analysis they vanish completely), with the seventh and higher harmonics being the same as in the simple square-wave modulator. This is a substantial gain in performance without too much design effort. The approach is widely used in commercial synchronous detectors.

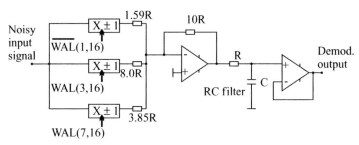

Figure 5.15 Walsh modulator. Each binary multiplier is driven by one Walsh function, with the scaling gains applied in an analog gain stage. The result is equivalent to multiplication by a sine wave approximation.

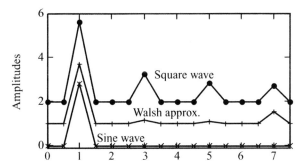

Figure 5.16 Fourier spectrum of the original sine wave, the Walsh function approximation, and the square wave approximation.

5.7   The Lock-in Amplifier

Commercial, general-purpose lock-in amplifiers provide the multiplication and integration functions of the basic synchronous detector and usually also provide the Walsh demodulator. Figure 5.17 shows a generic device. They are additionally endowed with many powerful and sometimes useful functions. In particular they often include:

- Current input front ends (similar to a transimpedance amplifier)
- Voltage input front ends (very high impedance)
- Differential inputs, for use in reducing ground loop problems
- Wide-range gain controls, allowing measurements from nanovolts to volts
- Input filters, low-pass, high-pass, bandpass, and 50/60 Hz notch
- Reference channels that accept a wide variety of waveforms (sinusoidal, pulsed, etc. and lock in to the correct frequency)
- $f/2f$ reference generators
- Wide-range and stable phase shifting (continuous and ±90°)
- Wide-range postdetection filter time-constants
- Different order postdemodulation filters (6, 12, 18 dB/octave)
- Two-phase (sine/cosine) detection.

The $2f$ reference generators are useful in situations where excitation of a system at frequency f produces primarily effects at $2f$. An example is a heating system, where an AC current drive I at a frequency of f hertz provides power proportional to I^2 and a temperature variation with much of its energy at $2f$ hertz. You would also use it for filament and gas discharge lamps, for example if you needed

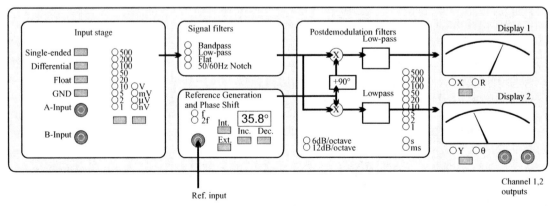

Figure 5.17 Generic lock-in amplifier, showing wide-range gain stage, reference generation, demodulation, and filtration.

a lock-in amplifier to detect office fluorescent lighting! The 50/60 Hz line voltage would be applied, suitably attenuated, to the lock-in reference input, the detected fluorescent light to the input. The dominant modulated light component is at 100/120 Hz since the light output when driven by AC is independent of the sign of the current and hence is at double the drive frequency.

As mentioned earlier, the newest lock-in amplifiers digitize the signal at an early stage and perform much of the processing numerically. Hence they tend to offer an even wider range of features. The distinction between these beautiful systems and vector voltmeters, spectrum analyzers, and scalar and vector analyzers is now becoming blurred. There is presumably still a market for analog instruments at the high-frequency end of laboratory measurements, but even this niche will be eroded as time goes by.

5.7.1 Phase shifting

The other crucial function block in all lock-in amplifiers is a phase shifter. We have seen that to obtain maximum output from the multiplier–integrator, it is necessary to align clock and input signals. There will always be some delay between the reference clock and the received modulated light. This is generally not due to time-of-flight effects, but to group delays in the photoreceiver. It is good practice to design the receiver to have adequate but not excessive bandwidth. This means that the square-wave modulated light is detected as a less-than-square electrical signal. The receiver rise time will be slower than that needed to accurately reproduce the square input light, and this slowing of the transitions is equivalent to a time-shift of the fundamental frequency which is detected in the demodulator. To compensate for this delay, the reference channel is provided with a wide-range, finely adjustable, and stable phase shifter that allows accurate alignment of reference and signal.

Where direct electronic modulation of the source is not possible, for example in some types of slow turn-on lasers and high thermal capacity sources, alternative intensity-modulators are needed. All the well-known optical modulator types may be used for this (e.g., Pöckels cells, acoustooptic modulators, waveguide and bulk electrooptic modulators, polarization modulators, liquid-crystal devices) as can mechanical modulators such as moving mirrors, moving fibers, and vibrating and rotating "choppers."

The classic chopper, a rotating disk with drilled holes or machined sectors, is still very useful and widely used (Fig. 5.18). As long as the beam transmitted through the sectors is small compared with the sector angular width, almost square-wave modulation of the light is possible. Where the beam is too large, the transmitted beam modulation is more trapezoidal. As we only detect the f_{mod} fundamental, this is of little consequence. A reference signal is provided by an optocoupler, magnetic Hall-effect device, capacitive sensor, etc. It is possible with this approach to vary the phase difference between transmitted beam and reference signal by moving the optocoupler azimuthally around the disk or equivalently by shifting the main beam. In practice both are usually fixed, with

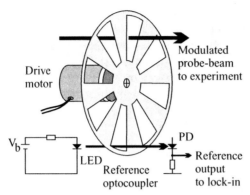

Figure 5.18 Chopper-disk used to modulate a continuous light source. By shifting the reference optocoupler azimuthally, the phase difference between reference and signal can be adjusted.

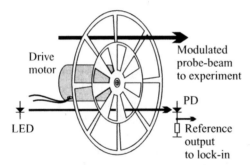

Figure 5.19 "Thin chopper" provides weak modulation at both even and odd harmonics of the fundamental. A separate reference channel allows use with a conventional lock-in amplifier.

the phase-alignment being carried out electronically in the lock-in amplifier. Problems with many motorized choppers are their poor frequency stability, leading to significant amplitude drifts when narrow bandwidths are used, and mechanical vibration. Even if this is quite weak, it can cause strong optical intensity modulation if laboratory hardware is insufficiently rigid. Precision mirror mounts and fiber-alignment systems are particularly open to this. The key characteristic is that the spurious intensity modulation is synchronous with the lock-in reference or its harmonics. Hence synchronous demodulation may not suppress these effects. Other types of mechanical beam modulators such as resonant tuning fork systems may perform better. Best of all are the electrooptic modulators without any moving parts whatsoever.

Usually we aim for 50/50 perfection in the chopper wheel, but this is not always the best approach. In cases where the beam power is important, such as when exposing photoresist or heating a substrate, the 50 percent power loss may be unacceptable. Here a *thin chopper* may be useful (Fig. 5.19), still imparting some modulation to the beam for detection, but allowing most of the beam to pass. The effective modulation frequency spectrum can be obtained from the Fourier transform of the transmitted intensity. The energy at the fundamental chopper frequency will be greatly reduced and accompanied by strong even and odd harmonics. This may make life difficult for the lock-in reference channel, so a square-wave fundamental reference may be needed which reads a different part of the wheel from the main beam. More complex choppers consisting of such coded disks are useful in some experiments.

5.7.2 Setting the detection time constant

The detection time constant can be adjusted on commercial lock-ins over a wide range, typically from 1 ms to 10 s. The detection bandwidth is approximately 2× the reciprocal of the time constant, and the narrower the bandwidth the less is the detected noise power. However, the narrower the bandwidth, the slower will be the response time. As usual, a compromise must be reached. A further consideration is the presence of interference signals. The discrete interfering signal shown in Fig. 5.20 will be detected if it lies within the passband of the shifted low-pass response. As the reference clock and the interfering signal are unlikely to be phase-coherent, the magnitude of the detected response will vary with time, leading to strange, beating signal variations. To reduce the spurious signal magnitude the modulation frequency can be shifted slightly away from the interference, or the passband can be narrowed, or the slope of the filter cutoff can be increased. Most commercial lock-ins allow at least "1-pole" (–6 dB/octave, –20 dB/decade) or "2-pole" (–12 dB/octave, –40 dB/decade) responses, and higher order postdemodulation filters can be useful in some circumstances. As we have recommended several times, it pays to have a good look at what interfering frequencies are present using a spectrum analyzer.

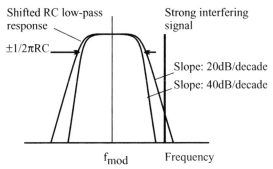

Figure 5.20 Suppression of a strong interfering signal close to the modulation frequency can be obtained by narrowing the response (increasing the filter time-constant), by increasing the low-pass slope, or by increasing the separation from f_{mod}.

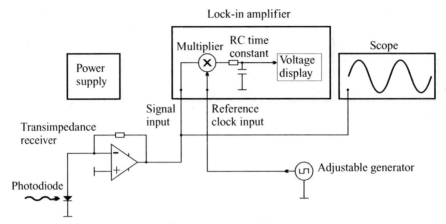

Figure 5.21 With an adjustable reference frequency, different frequency components present in the input can be detected. The system is a slow spectrum analyzer.

5.7.3 General detection

We need now to ask what is the lock-in's response to a general, nonsynchronous signal. Figure 5.21 shows a lock-in with an adjustable frequency generator driving the reference channel. Just as with an interference signal, the signal components falling within the postdemodulation low-pass filter will be detected, depending on their magnitude and their phase with respect to the reference. The phase will be arbitrary, so with a single channel lock-in the detected output will be variable. A two-channel lock-in however, which has identical product detectors driven by a sine and cosine reference, is able to determine the magnitude of the signal without knowing its phase. Hence slowly sweeping the reference generator allows the amplitude spectrum of the input signal to be measured.

5.7.4 Wire-free operation

This principle can be very usefully applied to large-dimension optical measurements. Figure 5.22 shows a modulated light transmission system without a common reference channel. Instead, both source transmitter and receiver lock-in are driven by different generators. Their frequencies are different, but very similar. In this example a divided-down 3 MHz crystal oscillator has been used for the remote source. A 32 kHz watch crystal could have been used. Such crystals are low-cost, very small, and are specified with an initial tolerance of ±20 ppm. This represents only ±0.16 Hz variation at about 8 kHz.

If you trigger an oscilloscope with the lock-in's reference clock, the optical signal from the remote transmitter will be seen to drift slowly across the scope screen. Warming one crystal with a touch of a soldering iron will change the drift velocity, increasing or decreasing the phase advance. As long as the pass-band of the lock-in is set to somewhat greater than 0.16 Hz, the optical signal

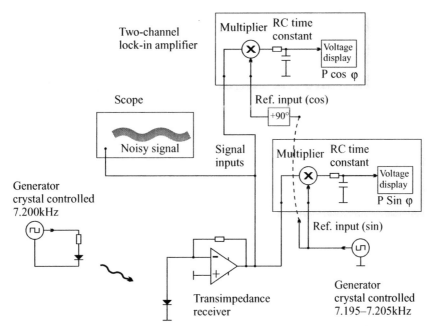

Figure 5.22 The lock-in reference signal need not come from the source modulator. As long as the frequency difference is small enough to remain within the filter bandwidth, separate oscillators can be used. However, *two-phase* detection will be needed to measure the signal amplitude independent of drifting phase.

will be correctly detected, although a single-channel lock-in will show periodic positive and negative output excursions as the phase varies. Taking the square root of the sum of the squared outputs from a two-channel lock-in, driven by sine and cosine references, we obtain a stable determination of the magnitude of the detected signal, without signal fading due to the slow phase variations. The two-channel lock-in provides the sum-of-squares computation via its magnitude/phase setting (sometimes labelled "$R\theta$"). Let's look at a nonsynchronous experiment in some more detail.

5.7.5 TRY IT! Non-phase-synchronous signals

The lock-in lends itself supremely to experimentation. There is sufficient complexity to test many different aspects of signal processing, but it has sufficient precision to make it all convincingly close to theory. Set up two oscillators, one being preferably crystal controlled and the other variable. The crystal oscillator could be made with 32 kHz watch crystals and two divided by two flip-flops (see Fig. 6.8). I had a 3.6864 MHz crystal divided down to about 7.2 kHz using an HC4060 chip. Use this to drive the reference input of your lock-in amplifier.

The other oscillator should be tunable around the reference frequency and as stable as possible. If you have a nice synthesized source, settable to a fraction of a hertz, all the better, but this TRY IT! can be done even with a dial-tuned analog oscillator with a little care. To be proper about this, you could drive an LED from the oscillator and connect the detected light to the lock-in input. If you are short of time, just connect the attenuated variable oscillator to the lock-in input. Fixed and reference oscillators may be exchanged, as in Fig. 5.22. It is also worth displaying both reference and input signals on a scope, triggered for example from the lock-in reference. The two signals are not phase-coherent, but with careful adjustment of the variable oscillator you should be able to get them within 1 Hz of each other or better. On the scope, one waveform will drift slowly past the other.

Set the lock-in time constant to $TC = 1$ ms, make sure the reference input is locked, and wind up the input gain to give half of full-scale deflection. The needle of the amplitude display (analog displays are *much* better for this experiment) should move back an forth between positive and negative peak values. If the input oscillator is really well aligned with the crystal reference, the needle may hardly move. At this point the reference and input frequencies match to much better than 1 Hz. With my digitally adjustable sine-wave input at 7.2009 kHz, the needle was barely drifting. Now offset the frequency as little as possible. If the lock-in is in two-channel *XY* mode, the two outputs will increase and decrease periodically as the phase between reference and input varies. Time the needle oscillation period with a stopwatch, and measure the input frequency if you can using a digital frequency meter. Adjust the frequency a little further and measure again. I did this a few times, plotting the graph of Fig. 5.23. For each setting, the detected amplitude display oscillates at the difference frequency. With the stopwatch and measuring ten cycles it was not hard to get down to a 0.35 s oscillation period. The display is just showing the beat frequency obtained by mixing

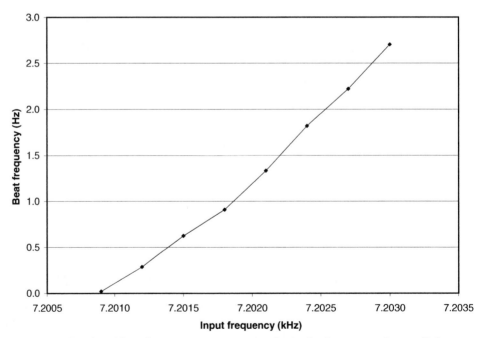

Figure 5.23 Results of beat-frequency measurements obtained using a crystal controlled source modulator and a finely variable reference clock. The beat was easily visible on the lock-in's analog display.

reference and input. The sum frequency $f_{ref} + f_{input}$ is filtered out by the lock-in while $f_{ref} - f_{input}$ is shown on the display (after filtering at the chosen time constant).

Now readjust the frequency to be a few hertz from the reference. The display will show the detected amplitude, oscillating fast between positive and negative extremes. Switch the display to show magnitude (R) of magnitude/phase ($R\theta$) mode. The display should stop oscillating, and give a more or less constant value. The two 90°-phased detectors are now added as sums of squares, giving the amplitude with the phase variation removed. Now increase TC to 1 s, and the signal should drop almost to nothing. This is because the acceptance bandwidth of the synchronous filter has become narrower (about ±0.16 Hz) than the separation between reference and input. Adjust the frequency carefully to line them up, and the signal appears again.

Now set $TC = 10$ ms, and adjust the input frequency to give the peak detected amplitude in $R\theta$ mode. Note the amplitude, and then adjust the frequency above and below the center to reduce the amplitude to half its peak value. In 6 db/octave filter mode I got 7174 Hz and 7228 Hz (54 Hz), and with the 12 dB/octave filter I obtained 7186 Hz, 7215 Hz (29 Hz). This gives an idea of the detection bandwidth of the synchronous filter in the two modes. Clearly, the 12 db/octave setting is about half as wide as the single-pole response.

Go back to the $TC = 1$ ms setting and XY display mode with an input frequency about 15 Hz above or below the reference. The amplitude display needle should be vibrating rapidly around the zero reading. It is clearly not able to follow the 15 Hz beat signal with full amplitude, but at least you can see it trying! (With a digital display all you get is fuzz.) If your lock-in has a connector output for the display voltage, you should be able to see the sinusoidal beat signal on a scope, which doesn't have any problem following the 15 Hz beat. Offset by 50 Hz, and the scope will show the 50 Hz beat, unless of course you attenuate it by increasing the filter time constant to 10 ms or longer.

Depending on what equipment you have, these experiments can be done in different ways. Instead of the external variable oscillator, you may be able to use the internal reference oscillator, if the lock-in has one. If not, two watch-crystal oscillators of nominally identical frequencies can be used. You can usually pull one of the oscillators either by loading with a small (100 pF) variable capacitor or applying brief jabs of a soldering iron to the crystal package to sweep one past the other with temperature tuning! Just be careful that there is no mutual electrical interference between the two generators, or they might lock together. All these experiments make great hands-on class demonstrations.

5.8 Spread Spectrum References

You may have noticed that in Fig. 5.14, one of the waveforms W(7,16) is not "square," but is nevertheless applied to a binary multiplier in Fig. 5.15. This suggests that there is nothing sacred about sine waves or single-frequency square waves and multipliers. More generalized source modulation and synchronous detection with an identical waveform works, and can be useful. In Fig. 5.6 we saw the importance of being able to choose our modulation frequency to avoid strong interfering signals. But what if the locations of these are unknown, or change from day to day, or are even being actively adjusted to coincide with your modulation and hence ruin your experiment. This is a scenario of "electronic warfare." By choosing a modulation waveform that is not centered on a fixed frequency, you may be able to gain in two ways. First, your signal energy will be spread out over a wider band, so that a fixed interference will have less of an effect on your demodulated output. Second, as the energy is spread out in frequency, the amplitude of individual components will decrease, making it harder for the "enemy" to find and target your signal. In the limit, with a noise-like modulation, your modulation may look just like natural noise and be invisible. This is one goal of "spread-spectrum techniques." While it would be a sad day if research students were actively trying to spoil each other's experiments with electronic warfare techniques, the experiments are didactic.

In fact, most of the lock-in functions described in this chapter have used a spread spectrum. We know that use of a square-wave reference clock opens up a multiplicity of passbands at harmonic frequencies, and if our receiver bandwidth is high enough to pass a few of them, we have a matched filter for a distributed frequency signal. Another of the useful waveforms for such spread-spectrum techniques is the Walsh function we looked at earlier. Figure 5.24 shows Walsh function modulation and synchronous Walsh demodulation applied to an optical channel. Figure 5.25a shows a spectrum analysis of the Walsh function WAL(7,8). It is clear that the modulation energy has been spread over a much wider range than the closest square wave of Fig. 25b. For this, the Walsh function generator was formed simply by hard-coding the waveforms into a PIC microprocessor and then reading them out in turn for output to a digital pin. The bit-period here is 1.04 ms, so that the highest fundamental frequency present is about 480 Hz. WAL(5, θ) is equivalent to a 480 Hz square wave, and WAL(1, θ) has a nonuniform mark-to-space ratio. It can be seen that there are many more spectral lines present with the nonuniform waveform, which can help to suppress interference.

In general, these non-simply-periodic signals cannot be used with a laboratory lock-in because the internal reference channel circuitry tries to lock its internal oscillator to the mean input frequency (that's the only "lock-in" thing about them). They do not simply perform the multiplication requested. The cir-

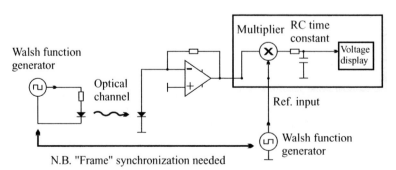

Figure 5.24 A Walsh lock-in is also possible, but now frame-synchronization, not just bit-synchronization, is necessary.

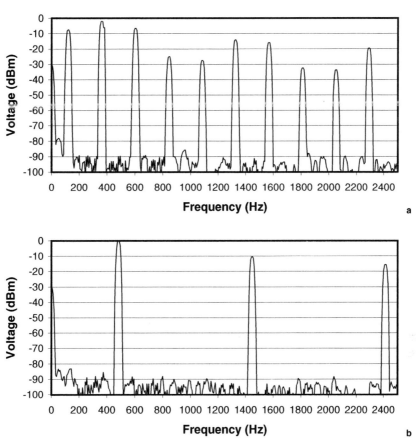

Figure 5.25 The Walsh waveform shown (a) is spread over a wider range of frequencies than the equivalent squarewave (b).

cuits shown earlier in the chapter can be used. I have used the AD630 IC driven by this binary waveform, which seemed to work fine for demodulation of the Walsh-coded signals. One disadvantage and even a reason for little use of Walsh-coding in general measurement is the requirement for the "frame-synchronization" shown in Fig. 5.24, rather than just the cycle-synchronization of a conventional lock-in. In the extreme case of spectrum spreading the modulating signal is white noise, or at least an approximation by pseudo-random binary sequences (PRBS). Some examples of noise generators are given in Chap. 3.

Synchronous detection has been shown to be a powerful technique to optimize optical measurements by allowing choice of a quiet region of the spectrum and by restricting measurement bandwidth as much as necessary. The most common expression of this is the commercial lock-in amplifier. There are many other signal extraction tools, such as autocorrelators, boxcar integrators, and phase-locked loops, as well as a further palette of software-based techniques such as Fourier analysis, Kalman filtering, and maximum-likelihood techniques. However, none seems to have quite the same characteristics of simple analysis and straightforward hardware realizations, so perhaps this is the reason for its great appeal. On the other hand it may be just chance, in the same way that Fourier analysis is so ubiquitous in science teaching, while the use of other orthonormal sets is so rare.

Useful Electronic Circuits and Construction Techniques to Get You Going

6.1 Introduction

So far we have covered some of the theoretical basics involved in making good optical laboratory measurements and in the electronic circuitry needed. We know a little about sources and detectors, their noise, and the added noise of amplifiers and about the beauty and power of modulation and synchronous detection. Circuit diagrams, calculations, and simulations are all very necessary, but whatever measurements you want to do, at some point you must put the theory into practice and build something! Sure, things will go wrong and not work as you hoped, so it is important to learn about the limitations and practical problems of electronics and optoelectronics in this environment, debugging the problems and figuring out what caused them. In my experience this means making *lots* of circuits, so you don't want to waste time searching for circuit fragments to do the job or actually constructing them. In general, the same needs and problems crop up time and again. Hence it is useful to have a collection, initially just designs but over time a physical collection, of the key circuits. Source drivers, synchronous detectors, clock oscillators and so on will be very useful in quickly assembling experiments; it is the flux of these experiments, properly evaluated, that generates experience. Later in this chapter we give a small collection of the "electronic clichés" you will need, together with some suggestions for actual components. Generally the circuits are rather trivial, but not if you need to search hours for them or spend time getting them to work.

6.2 Circuit Prototyping and PCBs

There are many different ways to rapidly prototype small circuits. For instance, the hobby catalogs offer "pluggable" circuit boards for through-hole components aimed at this need. With a hundred or so holes arranged in an *XY*-array in a plastic brick, they have all the *X* axis rows connected internally. The idea is that you insert the components and bridging wires more or less in the *Y* direction, with the *X* connections taken care of internally. If you use these, it is important to use solid-core connecting wires to allow easy insertion. Stranded wire splays and either won't go into the holes or goes in and makes connections you don't want.

However, the plastic blocks are not usually screened, and the large conductor strips attached to every point in the circuit can make for odd performance. Wrapping aluminum foil around the block and grounding it with an alligator clip can help, but I have never been very happy with this solution. On many occasions I used to forget about the *X*-direction connections, with disastrous results. They work better with digital logic ICs than with the variable packages of analog components.

I have similar small accidents with copper strip board. This is insulating material with a dense array of punched holes (usually on 0.1-inch centers), and parallel strips of laminated copper foil, protected by a solderable lacquer. The connections must be soldered for this. The ideas of using the board to make the *X*-direction connections, leaving the *Y* direction for component leads and wire bridges is similar to that of pluggable boards. At least here the hole and component density is high enough to be useful, and once soldered the components remain connected. You also have the option to sever the copper strips as needed using a few turns of a handheld drill bit or the tool supplied for the job. It is a big advantage to plan and draw out the circuit completely before starting to solder. Even coarse graph paper or a word processor with a fixed character pitch font is suitable for this. As with pluggable boards, I generally forget to sever a few lines, with results that are at best hard to understand. Reaming out the little copper areas to isolate different regions of copper strip is also fraught with risk; it's easy to leave a few microscopic shards of copper to short out power lines and clamp logic gates to one state. A good inspection under a lensed desk lamp or low-magnification stereo microscope is beneficial. Most layout problems are visible.

Years ago I used to spend hundreds of hours carefully laying out professional printed circuit boards on a CAD system. These had to be photo-plotted on plastic film, and then sent for transferring onto copper laminate board for subsequent etching. This took several days but only took a couple of minutes to discover that the circuit was seriously flawed. The throughput of this approach can be limiting.

The approach that has lasted best with me is the use of *blob board*. This is single-sided, copper-laminated composite board material, just like the strip boards mentioned above, one of the family of Veroboard products. However, instead of copper strips, this one has a predrilled array of isolated square copper pads on 0.1-inch centers. In this case you have to make all the connections yourself. With a small soldering iron, flux-cored solder, and a little practice, it is possible to quickly "draw" circuits in solder onto this blob board, bridging between the copper islands.

6.3   Basic Receiver Layout

Let's look at a simple concrete example, the basic transimpedance amp (Fig. 6.1). It is an opamp design using a single 8-pin DIP amplifier package. A discrete transimpedance capacitance has been included to roll off the transimpedance at high frequencies for stability and to reduce gain-peaking. We are using symmetrical power rails, although the output will be only positive-going with the photodiode as drawn. I have included small power rail decoupling capacitors (e.g., 100 nF) which should be connected as close as possible to the IC. You might get away without them, but the 51 Ω resistors help to form a low-pass filter to suppress injected high-frequency noise and feedback through the power rails. The opamp's power-supply rejection-ratio is large but not infinite and degrades with frequency. It is easier if you do it every time. The 51 Ω value is chosen to drop not more than a couple of hundred millivolts at the peak opamp supply current. The resistors also make layout easier as they add an extra degree of topological freedom to cross over other conductors. I try to use the same lead spacing for all similar component connections, not changing the spacing to fit the design (0.2 inch for the tiniest $\frac{1}{8}$W metal film resistors, 0.4 inch for standard $\frac{1}{4}$W resistors, and 0.1 and 0.2 inch for capacitors).

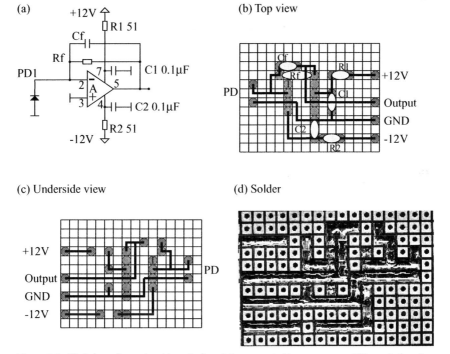

Figure 6.1   Blob-board construction. A circuit is converted to a square-grid layout showing all components and offboard connections. The underside view allows easy drawing of the circuit wiring in solder on isolated copper islands.

Construction starts with your circuit diagram, making sure that *all* components are present, including power supplies, decoupling capacitors, and offboard connections, and that all IC pin numbers are correct. Draw the components in pencil on coarse graph paper. I prefer something on a grid of about 5 to 6 mm. This can be bought through office supply and art shops. If it is a problem to find, print something equivalent from the PC. Each square represents the 0.1 inch of the blob board. It's not 1:1, but neither is it a problem to follow from paper to board. The aim is to lay out the component positions to optimize performance. Critical paths such as the feedback network and the photodiode connection should be as compact as possible to minimize unwanted pickup. We have seen how dependent performance can be on a few picofarads of extra capacitance here. Component connections are drawn onto the coarse graph paper using strict north, south, east, west routing. It might take a couple of tries to get a neat layout throughout. With a little luck, the components themselves can provide most of the flexibility needed, avoiding all but a few wire bridges. These can be of bare tinned copper wire for rectilinear runs on top of the board, or fine solid-core insulated wire for connections underneath. Tefzel or Kynar 30 AWG (0.25 mm) insulated wire-wrap wire is perfect for this, easily stripped with the right tool and proof against overheating by the soldering iron. For offboard connections you can use 1-mm-diameter pushpins ("Veropins") made for these boards, with soldered wire connections. Tiny screw terminals with connections on 0.1-inch centers are also available, which work well with blob board. Make sure you use sufficient pushpins to act as test points for all important signals and a convenient ground pin or wire-loop for your scope probe ground. Changes and errors are quickly corrected with an eraser. Figure 6.1b shows the design layout. On this layer I have the components, wire connections, and all the labels. This view will be used for debugging.

Once the layout is finished, I make a hand copy onto tracing paper (or matt polycarbonate drawing sheet) again with a pencil. This is very quickly done, taking only a minute or two. Only the connections are drawn on this layer not the component bodies. You can avoid one step by drawing straight onto translucent graph paper. Art shops sell this too, or you can print your own. The only problem is that the underside view still includes all the component bodies, which is too much information. Hence I prefer the two-sheet approach.

Working from the top view and following the layout, the components are now inserted into the blob board exactly as in the paper layout. Count squares to make sure everything looks right. Small placement errors can be trivial or disastrous, depending on the local topology. Clamp the components with a cloth or soft sponge, flip the board over, and solder the leads into place. Just one connection per component will do. A soldering iron with a round 1.5-mm-diameter tip is about right, allowing one pad to be heated without touching its neighbor. Check that everything is held in place, and correct any components that are falling out of the board or leaning over. Complete all the component soldering and cut excess component leads flush with the solder bumps using sharp sidecutters. Now the "wiring" can begin.

Turn the board and the tracing layout over to view the underside (Fig. 6.1c,d). The components will still be obvious by comparison with the traced view, if it hasn't been weeks since you made the layout. Following the traced wiring, you now need to "draw" the wiring with solder. This is a rapid process. The board should be held firmly in a small vice or clamp, as you will need to apply some pressure from the soldering iron to get fast, thermally conductive connections to the pads. A small rotational oscillation of the iron in your fingers scrubs the tip against the pad and greatly improves heat transfer. First put a solder blob onto each square pad along the wiring route, using a touch of the hot iron and then a couple of millimeters of flux-cored solder. Once all the solder bumps are in place, they can be joined together. This involves melting the second bump in a row, dragging the molten solder over the insulating gap back to the first solid bump, adding a small additional amount of fluxed solder, and then quickly removing the iron. If this is judged correctly, the molten blob will melt into the previous solid blob, form a good connection, and promptly solidify. Solid solder will then have bridged the gap, and the second and third blobs can be joined in the same way. If too much heat is applied, usually by performing the backward sweep too slowly, both blobs will melt and the high surface tension will separate the solder into two unconnected islands. Immediately trying again will often frustratingly just add to the size of the two molten blobs, because the temperature of the whole area is now too high. If this happens blow on the board or wait ten seconds for everything to cool down and try again. This is all easier to do than to describe, although it does take a little practice. Figure 6.1d shows the final result.

Once you have the touch, this drawing in solder process can be very efficient and extremely fast and rarely results in misconnected circuits. You have, after all, sorted out all the topology beforehand on paper. Nevertheless, a final check of the physical board against the underside paper view is recommended, again using the magnifier or stereo microscope.

When all wiring is complete, you will have used a lot of solder, so much of the circuit will be covered in excess flux; this must be removed. The commercial aqueous flux cleaners are ideal for this. Immerse the whole board in the cleaner for five minutes, and then scrub the board underside with an old toothbrush. Finally, rinse the board under running tap water until all trace of a soapy feel has gone. If you have any, rinse also in deionized water, dry the board on laboratory paper or a paper towel, and then place in a warm airing cupboard or under a tepid hair dryer to dry. As we discussed in Chap. 3 failure to remove the flux can lead to odd performance in detector circuits, due to its significant conductivity. A few components, notably some piezo sounders, batteries, switches, and unsealed relays, should not be immersed. They will be added after washing and drying.

The act of laying out on paper helps to make the layout perfect, so that I rarely need to make any corrections. Gross errors can be corrected using copper braid or a "solder-sucker" to aspirate solder along a line of connection bumps. These work very well on the single-sided, nonplated hole boards. Similarly, components can be exchanged by sucking up excess solder around the leads and pulling out the component. If extensive modification needs to be done in sensitive areas, such as around the inputs and feedback resistors in transimpedance amplifiers, flux cleaning may need to be done again. If you change the layout, change the paper drawing to match. If you need to review the design in the future it should correspond with what you see.

The main advantage of this well-practiced approach is its speed. Small circuits can be drawn, laid out, and soldered in an hour or two. If you keep collections of designs for preexisting transimpedance amplifiers, synchronous detectors, modulators, power supplies, filters, etc., they can be easily reused with very little new layout work. With simple circuits you may have a sufficiently good mental picture of the mirror-image underside view to avoid the transparent paper step. The only remaining task is to file away your design so that it can be reused in the future and perhaps to label the board's external connections. A word processor set to print small labels in 4- or 6-point type for gluing onto the board is about right. A digital photograph of the board can be very useful; scan the completed board, top view and underside, on an office flatbed color scanner. The depth of field is usually adequate.

If you need a couple of identical simple boards, the blob board method is still the best approach. Once you need more, and the performance has already been verified, this approach represents too much work and a conventional PCB is probably a better approach. Once the graph paper layout has been designed and tested, it is usually straightforward to transfer to a PCB layout. Almost all of the layout thinking and optimization has been done already. The main difference is that the blob board doesn't allow tracks to be placed between IC pins, so the PCB can be a little denser.

6.4 LED Drive and Modulation

It may seem a little trivial to discuss how to drive an LED, but it is not uncommon for users to treat them like a lightbulb, connect them to a battery, be impressed at how bright they are, but disappointed in their short lifetime! The problem is that they are diodes, and Shockley said that the current flowing through them varies roughly exponentially with the voltage across them. Hence although you *can* drive them from a low-impedance voltage source such as a DC power supply, even small increases in supply voltage will lead to large and possibly destructive increases in current. It is therefore much more pleasant to use a higher-impedance source that defines the current and allows the diode voltage take on whatever value it needs.

To use the LED you need to know the maximum forward current that it can bear (typically 20 mA for a common visible LED, perhaps 200 mA for a high-power infrared device) and the maximum reverse voltage it can handle without breaking down and passing damaging currents (typically only a few volts). You get these figures from the data sheet. Then we can choose the voltage source V_b and series resistor. There is a lot of leeway. A low voltage and small resistor works as well as a high voltage and a high resistor. However, the higher the voltage, the more like a pure current source it looks and the more stable will be the current with temperature variations in the LED. In Fig. 6.2a the current $I = (V_b - V_f)/R$. Unfortunately, the diode voltage V_f is also a function of current, but not much of one, just logarithmic. Hence a reasonable approach, since you probably don't know the details of the IV characteristic of the cheap LED, is to take any available 5V or 12V DC supply or battery, assume a forward LED voltage of 1.5V, an operating current of 10 mA and calculate the load resistor. Power it up, measure the LED voltage, and adjust the resistor if necessary. It's unlikely that you will be far wrong. If the LED is more costly or you don't have a replacement, you might start with a ten times larger resistor or wind up the supply voltage slowly, monitoring the current until you are sure of the polarity, the characteristics, and hence the necessary design.

Connecting to a laboratory signal generator to modulate the LED is hardly any more complex (Fig. 6.2b). Use the peak generator voltage instead of V_b in your calculations and proceed as above. Unless the LED has a metal can which you would like to ground, you might want to swap the positions of LED and load resistor as shown. This makes it easier to monitor the LED current via the voltage on the load using a grounded oscilloscope. If the generator is bipolar, one half-cycle could take the LED into breakdown. If you suspect this, limit the

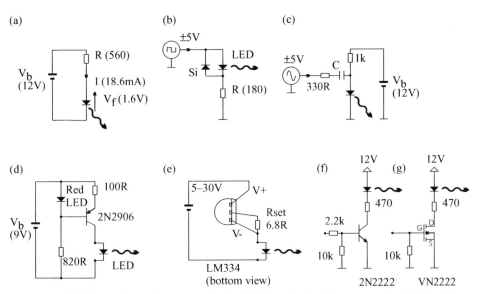

Figure 6.2 LED drive circuit fragments. As the LED is a diode, it is more comfortable to define the LED current and let the voltage take on whatever value it needs. (b) and (c) show square wave and sine wave modulation schemes. (d), (e) are simple DC current generators. If more current is needed than an opamp or CMOS logic gate can supply, use a bipolar transistor or FET buffer (f,g).

LED reverse current or voltage by connecting a silicon diode either in series or in reverse parallel, respectively. The load resistor should always be large enough to avoid exceeding the current output capability of the generator.

This works fine with square-wave generators and on/off LED output. If you want instead to sine-wave modulate the output, you will need to bias the LED with a DC current first. One way is to use a DC supply and resistor to give about half the maximum allowed LED forward current, and AC couple the bipolar sine-wave generator to the LED, as shown in Fig. 6.2c. The AC generator can then add its current up to the LED maximum and reduce it to zero. As before, make sure the generator can drive the series resistor load without damage. The capacitive coupling and series resistor lead to a high-pass characteristic, so C has to be large enough not to attenuate the frequency you want to use. Choose $RC \gg 1/f_{mod}$ and use a nonpolarized capacitor. Using a little care in setting up the other components, the capacitor can be dispensed with altogether. To set up the modulation amplitude, observe the LED output on your receiver. Too much modulation current or too little bias will lead to a clipped sine-wave output.

If you use a lot of LEDs, in routine testing it is useful to have an active current generator set up for about 10 mA. This can be a single-transistor circuit as in Fig. 6.2d. The red LED acts as an indicator and also provides a relatively constant voltage of about 1.6 V to the transistor base. The base-emitter junction loses another 0.6 V, to give 1 V across the emitter resistor. One volt and 100 Ω is 10 mA, which also flows through the collector and LED. Whatever the LED and even if the LED terminals are shorted together, the current remains near to 10 mA, so the circuit is quite robust. Last, there are even simpler integrated circuit current generators available such as the LM334 (Fig. 6.2e). This is a high-precision current regulator that can deliver up to about 10 mA. Both Figs. 6.2d and e only need a small battery with 6 V or 9 V for long-life operation.

When you want to drive the LED from low power circuitry, including opamps and CMOS logic, their current capability may be a bit marginal. In that case use a bipolar or FET transistor used as a switch, as in Figs. 6.2f and g. Any small npn-transistor will be fine, such as a BC548, 2N3904 or ZTX450. You can even avoid wiring the two resistors by choosing the new "digital transistors". Infineon does a good range, such as the BCR108, BCR112 which can sink 100 mA. Rohm's range (e.g., DTC114EKA) is very similar. These are surface-mount devices. Arrays of small switching transistors in a single package are also convenient. Intersil's CA3081E contains seven common-emitter npn-transistors in a DIP16 package. For the FET any small n-channel enhancement-mode MOSFET with gate threshold voltage less than about 3 V is convenient. The VN2222LL is widely available, and for higher currents Zetex does a good range such as the ZVN2106A series. These have a channel on-resistance less than 2 Ω and can handle 450 mA continuously. As they are operated as switches, both drivers need a series resistor to limit LED current.

It is not always necessary to drive the LED at or below its nominal current. The main limitations to drive come from temperature rise of the LED chip,

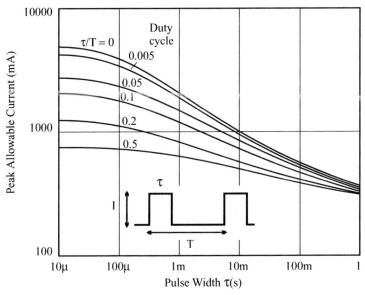

Figure 6.3 A reduced duty-cycle pulsed drive allows LEDs to be driven well beyond their normal maximum current to obtain higher peak outputs.

rather than the instantaneous current that is being applied. So, if the current is applied only for a short time, the LED has no time to heat up and much higher currents can often be applied. The technique of applying higher currents in short pulses is widely used with infrared diodes used in remote controls and in visible multiplexed displays. In addition, physiological effects increase the apparent brightness of visible displays operated in this way, so efficiency improves too. The impulse loading information is normally available as a family of curves showing the allowable peak current versus the pulse width τ, for a range of duty cycles τ/T. Figure 6.3 shows an example, similar to the Siemens LD242 infrared LED. Typically, with a $100\,\mu s$ pulse and 1 percent duty-cycle the current could be ten times the DC maximum value.

6.5 Laser Drive and Modulation

Although laser diodes are in many ways similar to LEDs, with the same materials, electrical diode characteristic, current requirements, power dissipation, and optical power output, their operation is much more complex. This is because in addition to specified limits on drive current and reverse voltage, they have a maximum permissible optical power output. If this is exceeded, the very high power density at the laser facets is likely to degrade or completely destroy the laser. This process can happen in microseconds or less. Hence much of the design of laser drivers is concerned with avoiding even transient current drive overloads.

They also exhibit a strong nonlinear output/current characteristic, in which an increase in drive current initially produces little light output, but above a certain threshold current the output increases rapidly (Fig. 6.4). Even simple biasing is difficult, and it is not advisable to just connect to a variable current source and hope for the best, as with an LED. For example the drive current "Bias 1" in Fig. 6.4 may initially give the desired output, here about 0.75 mW. However, as soon as the laser is powered up, it also warms up, which can shift the threshold characteristic to the higher current characteristic shown. This may completely switch off the laser output by taking it below threshold. If the current is instead raised a few milliamps to give the increased output (Bias 2) at the higher operating temperature, and the ambient temperature drops a few degrees, the output power may easily climb to a value exceeding the specified optical power limit, causing degradation or damage. These characteristics are contained in the IV (current-voltage) and PI (optical power output-current) characteristic curves, which are supplied with the laser diodes, either individually or as family characteristics.

To deal with this problem of output power control, the great majority of low-power laser diode packages have three pins, connected to the laser itself and also to a monitor diode picking up light from the laser's rear facet. The photocurrent from this monitor diode varies approximately linearly with the laser's front facet output, and it is this signal that must be controlled to limit output power overloads. The high collection efficiency and hence high photocurrent

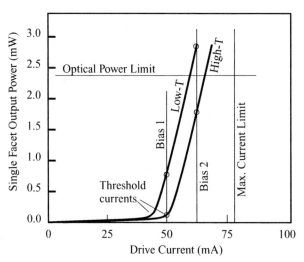

Figure 6.4 Laser diodes exhibit a threshold characteristic. Normally we bias just above threshold and modulate with current pulses above. There are both maximum current and maximum power limits for each device and the threshold current, in particular, is temperature dependent. Correct biasing is simplest with feedback control from a monitor photodiode.

detected means that sensitivity is not a problem (typically it is $100\,\mu A$ or more). The design with a common pin means that you will find four different connection permutations of commoned electrodes, which must be handled differently (Fig. 6.5). The "parallel" configurations in Figs. 6.5c and d are less convenient as they ideally need bipolar power supplies to power the laser and to extract the reverse-biased photodiode's photocurrent. The "antiparallel" configurations in Figs. 6.5a and b are easy to use with a single supply.

Typical electronic power control circuitry uses a feedback control system. As it is so important to avoid even short overshoot spikes, the controllers are usually set up as heavily damped PI (proportional, integral) types. These drive the laser with a signal that is the integral of the difference (error) between the set point and the actual photocurrent values. Figure 6.6 shows one possibility for the antiparallel form. The laser is driven from a high-impedance single-

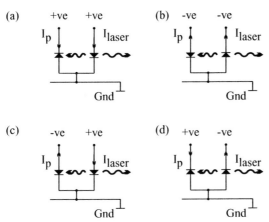

Figure 6.5 All four permutations of laser and monitor photodiode polarity are available. The "antiparallel" configurations are slightly more convenient to use with a single-polarity power supply.

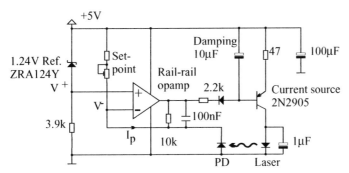

Figure 6.6 Example of a DC laser power-stabilization circuit.

transistor current source. Photocurrent flows through the variable resistor to give a voltage which is compared with a reference voltage. The set point is provided by a 1.24-V two-terminal band-gap reference IC (e.g., Zetex ZRA124Y or National Semiconductor LM385Z-1.2). The opamp adjusts its output to pass sufficient current through the laser to generate this photocurrent and reduce the voltage difference at its inputs to zero. If the photocurrent increases, V^- will drop and the opamp output will rise, cutting off the transistor and stabilizing the laser output.

Easy-to-use chips are also available for this function. For example the Sharp IRC301 is good for cathode-to-cathode package configurations. Due to the differences among lasers, their packaging polarity, and the widely differing monitor photodiode sensitivities, the reader should study the manufacturers' literature for circuits and precautions. That done, there is no reason why you shouldn't wire up your own. It is just important to think through the circuit operation, both during regulated operation and during power-up and switch-off. The use of a single supply rail, as in Fig. 6.6, avoids problems with the order in which bipolar rails appear. This is a frequent source of problems in home-made designs. Arrange for a slow turn-on set-point, and make sure that the main power supply is free from transient interference. The electromagnetic compatibility (EMC) regulations have spawned a huge variety of power-line filters, which can keep the worst interference from getting to your laser.

The most important aspect, however, is to evaluate and protect against dangers from visual damage. Although the output powers of laser and LEDs may be similar, the danger from lasers is much higher due their ability to be focused, either by instrument optics or the eye itself, onto a tiny spot of high-power density. Even a few milliwatts can cause permanent damage, especially with near infrared lasers which are almost invisible. If you can see the deep red light of a 780 nm laser, the density is probably far too high! Wear proper eye protection rated for the laser power and wavelength in use, even if it is uncomfortable. Check the protection wavelength every time you put them on—many different types look the same.

To modulate the laser's light output you normally bias the device just above the threshold knee and then apply small current pulses to drive up to the optical power limit. Hence two control loops are ideally needed to stabilize both the off- and the on-state powers. More often you will see just the off-state stabilized at threshold, with fixed-current modulation applied. This is acceptable, as the main variation with temperature is the threshold current increase. The slope above threshold is relatively constant, although it does decrease slightly with temperature. Some diode lasers show kinks in the PI characteristic above threshold, as different optical modes compete for the material's gain. This can make power stabilization at a high level difficult. Modulation from zero current is possible, but without prebiasing just above threshold, modulation speeds will be reduced by a significant turn-on delay.

Note that the power stabilization is primarily there to protect the laser from overload, not to give high precision in the output power. If you want high intensity measurement precision with a diode laser source, more care will be needed. Instead of using the laser's rear facet detector for stabilization, you could use the front facet output, for example using a small beam-splitter, and thereby remove one source of error. In practice it is easier to use the internal rear-facet detector for rough intensity stabilization, with an external precision detector for referencing. Note too that the laser stabilization circuits are just as applicable to light emitting diodes, given an independent detector. Techniques for power stabilization and referencing are described in Chap. 8.

6.6 Modulation Oscillators

In the great majority of optical measurement systems the light source will be modulated, and for LED-sourced systems the basic requirements are, well, basic. There is nothing very wrong with using the "555" timer IC in its astable multivibrator mode (Fig. 6.7). The simple circuit does not allow equal on and off times, but this can be remedied by moving R_B to the pin 7 connection, as on the right side of the figure, and restricting it to no greater than $\frac{1}{2}R_A$. Otherwise threshold will not be reached and the circuit will not oscillate. See the National Semiconductor Corp. data-books and application notes for details. They are all on the web site, but the printed information is more educational. This IC, one of the most successful and widely used of all time, is available in a variety of technologies. The original LM555 bipolar devices offered a powerful 200mA source and sink capability. This is sufficient to directly power many high-current LEDs. Newer CMOS versions are available with operating voltage down to 1V, much lower input and quiescent currents, and greatly reduced "crowbar" current spikes during the timer reset period. A better way to obtain accurate 50:50 mark-to-space (M/S) ratio square waves for synchronous detection use is to operate the oscillator depicted in Fig. 6.7 at double frequency and divide by two in a D-type flip-flop as below. This ensures accurate M/S even with highly asymmetric input waveforms.

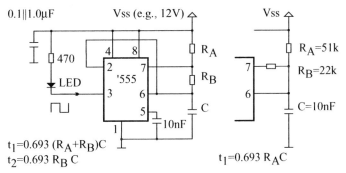

Figure 6.7 Modulation oscillators using the well-known 555 oscillators are common in optical circuitry. The right side shows one way to obtain 1:1 mark-to-space ratio drive.

The 555 oscillator can achieve a temperature drift of less than 150 ppm/°C, although typical capacitors will drift faster and degrade this figure. We saw in Chap. 5, however, that it is often helpful to be able to define the source modulation frequency with better precision than this. For example, to avoid the harmonics of fluorescent light interference we might want to set the frequency carefully between 4000 Hz and 4050 Hz and have it stay there with temperature changes. Or we might want two oscillators with very similar frequencies to operate a reference-free synchronous detection configuration. Even in noncritical applications, it is useful to know precisely what the modulation frequency is. In all these cases it will be easier to use a precision resonator oscillator, rather than an RC-oscillator design.

The cheapest route to initial frequency accuracy and temperature stability (typically ±20 ppm/°C) is the 32.768-kHz crystal used in most clocks and watches. These are available in several through-hole and surface-mount packages, including tiny 2 mm × 7 mm long cylinders. The most common way to excite all resonators is to connect them as a feedback element around a digital logic gate. Almost any inverting gate will do. This is the internal design of the Harris HA7210 used in Fig. 6.8, which has some other refinements to optimize performance. For example, current drain with the device is very low ($\approx 5\,\mu$A) and

(a) Divided Clock

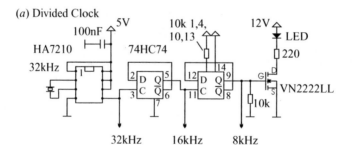

(b) Two-phase clock

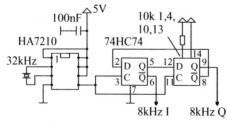

Figure 6.8 A better way to get 1:1 mark-to-space ratio drive is to generate any waveform and then divide by two with a D-type flip-flop (a). The 74HC74 series has two such elements per package, leading to compact and useful modulated sources. A different connection (b) gives division but also "quadrature" outputs.

includes switchable elements to optimize internal circuitry for low-frequency clock crystals or higher frequency resonators. No extra components are needed except for power supply decoupling, placed as close as possible to the chip. The only potential difficulty with this chip is the very high impedance at which the inputs work. Even low levels of leakage around the crystal connection pins will stop the circuit oscillating. This includes PCB leakage currents, so it is important that the board is carefully cleaned of flux residues and preferably guarded as recommended in the application guidelines. If the 32 kHz output is too fast for the application, division stages can easily be added in CMOS D-type logic as shown. The 74HC74 contains two dividers, getting you down to 8 kHz. The same components, connected differently, can provide 90° phase shifted outputs I and Q. Using CMOS logic, this module will operate any conventional LED using a single 3.3 V lithium cell and an FET or transistor buffer. Another useful IC is the 74HC4060, which combines a free inverting gate between pins 10 and 11 which can be used as an oscillator, and 14 stages of division, with most of the division stages made available on pins (Fig. 6.9). With some tweaking of the biasing components, this will work with a great variety of two-terminal resonators. The divider chain will get a 32 kHz clock crystal down to 2 Hz.

Ceramic resonators offer another approach to timing. For example, 455kHz resonators are available from Murata and others, which are economical and easy to drive. The HA7210 can again be used, or you can make your own circuit using simple gates. The Murata design notes are very helpful, and recommend a particular kind of logic gate. This is the CD4069UB series (Fig. 6.10). These gates are "unbuffered," which gives a reduced open loop gain compared with buffered gates, reducing the likelihood of oscillation at a harmonic of the design oscillation frequency. Sometimes it can be difficult to get any of these

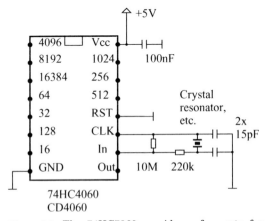

Figure 6.9 The 74HC7060 provides a free gate for exciting crystal resonators, together with a long divider chain for selection of lower frequencies.

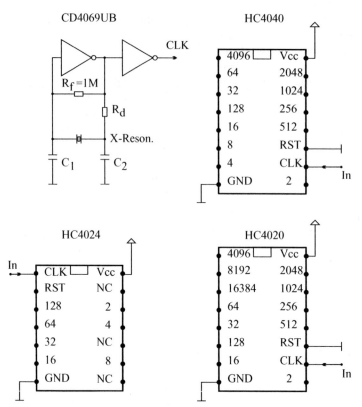

Figure 6.10 Ceramic resonators are best excited using "unbuffered" logic gates. You may need to tweak the capacitors to get it to oscillate reliably. Alternative divider ICs are also shown.

piezomechanical resonators to oscillate. This is down to the two loading capacitors C_1, C_2 shown in Fig. 6.10. With a good model of the resonator, they can be calculated and chosen analytically. In practice we are often reduced to tweaking to get them to work. Luckily, once tweaked, the design is usually transferable to other similar devices. Division of the fundamental frequency can simply be carried out in a divider chain, such as the 74HC4060 already mentioned or the related HC4020, HC4024, or HC4040.

Neither ceramic oscillators used for radio intermediate frequency filtering nor the watch crystal offer sufficient choice of frequencies for some applications. In this case, it is convenient to move to the 1 to 10 MHz region, where the choice of conventional quartz crystals is at its best. With correct choice of the loading capacitors, all the above circuits can be used. The HC4060 is a nice solution as it includes both the free gate to build an oscillator and a long divider chain to get down to the audio range. If you need even more flexibility, look at the frequency synthesizer chips.

Second only to the bias box in its continuous usefulness around the lab is a modulated LED source. The crystal-controlled audio-frequency transmitter (Fig. 6.8a) with a 3.6V lithium battery is about right for finding out whether your receiver has just died, checking the continuity of a fiber link, or testing the rise time of a new high-sensitivity detector circuit. The circuit is compact enough to be built into a flashlight housing or small diecast box. My recommendation is to make several and always keep a few hidden away, as they have a high diffusion coefficient! A good runner-up for testing receivers is a handheld IR remote control. As every TV, video, stereo, gas-fire, and air-conditioning system these days seems to have a remote control and they last longer than the equipment itself, there are usually surplus units available. The wavelength is typically 880 nm or 940 nm, although changing the LED for a different wavelength is not difficult. The slow code transmission can make scope triggering difficult.

6.7 Single Supply Receivers

As light detection and hence the output of a transimpedance amplifier is inherently *unidirectional*, it is attractive to use a single power supply voltage (Fig. 6.11). In particular, this could be same 5 V supply used with the clock oscillators, logic circuitry, and microprocessors of the rest of the system. This is possible, although there are some limitations compared with operation from symmetrical supplies. First, the photodiode is restricted to anode-to-ground polarity if a positive output is desired. As both inputs are at ground potential, we must also choose opamps whose inputs can operate there. If the receiver is used with modulated light or over a wide range of static illumination intensities, we need the output voltage range also to include ground. It is not necessary that the output voltage can swing to the positive rail, but it will help to

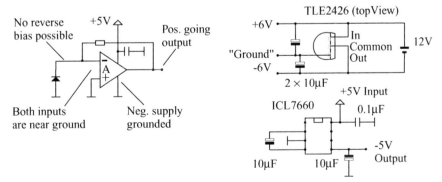

Figure 6.11 Single-supply operation of transimpedance amplifiers is possible, although performance may suffer. Bipolar supplies may be obtained with artificial ground generators (TLE2426) or a switched-capacitor voltage inverter (ICL7660 and equivalents).

give the best dynamic range. In a conventional opamp design the output voltage is limited to not closer than about 1.2 V (two diode-drops) from the positive rail, and this can be a big loss with a 3.3 or 5 V supply. Luckily, you aren't on your own in needing these characteristics; almost the whole electronics industry, driven by applications in mobile phones, digital cameras, personal digital assistants, MP3 music players, and notebook personal computers is pressing for new opamp designs that stress low-voltage, low-power, rail-to-rail inputs and outputs and operation from the same 3.3 or 5 V supply used for the digital circuitry. Hence there is a reasonable and rapidly expanding choice. The downside is that these requirements make design for high speed, low offset voltage, good output drive capability, etc. much more difficult. In transimpedance configuration some rail-to-rail opamps show severe distortion and instability with low output voltages. As your laboratory design is unlikely to be built or abandoned depending on the cost of a couple of chips, it is often worth just using split supplies, generating split supplies using a specially designed voltage regulator, or generating a low-current negative supply from the +5 V rail using an ICL7660 inverter. Figure 6.11 shows a few possibilities. These can be used to supply either the whole opamp or just the photodiode's reverse bias. Note that the oscillation frequency of the standard devices is fixed in the audio band and can give a lot of in-band interference which is difficult to screen perfectly. Alternative devices offer either much higher operating frequencies or adjustable frequencies, which can help to avoid the biggest problems.

6.8 Use of Audio Outputs

In many of the simple systems discussed we have modulated the source in the audio range, above the region of worst industrial noise, say 500 Hz to 20 kHz. Because of this, a receiver with an audible output is a very useful piece of equipment for optical system testing. This can be used to find out whether the transmitter LED is being modulated, or even if the invisible infrared LED is actually still alive. In debugging field installations, it may save lots of time and face. In multifrequency systems the human spectrum analyzer can even differentiate between sources and allows extremely effective detection in the presence of noise. You can of course make up a complete battery-powered transimpedance receiver, small audio amplifier chip such as the LM386, and a speaker or earphone. This is the proper approach for testing weak signals. However, even a trivial receiver can be useful.

6.8.1 TRY IT! Minimalist audio photo-receiver

A large area photodiode or small solar cell connected to a pair of headphones, personal audio player earphone, or even a piezobuzzer does a great job. Ideally you would like the most sensitive device you can find. Try to find an old high-impedance earphone; 600 Ω or 2 kΩ are best, but even a modern 16 Ω personal stereo earpiece works, albeit with lower sensitivity. Just connect it to a silicon photodiode. An economical BPW34 is fine, but so is almost anything else. I am lucky to have a twenty-year-old pair of high quality Sennheiser stereo headphones. These have an impedance of 2 kΩ each, so in series they work very well. Listen close-up to a few modulated optical sources: the PC monitor, TV screen, TV remote control, fluorescent room lights, or your 8 kHz crystal oscillator and LED driver. You should have no difficulty hearing them all. Desktop 100/120 Hz room lighting is a bit faint, but that is more a limitation of your hearing's low-frequency sensitivity, not the detector's. Reverse biasing the photodiode with a small battery improves performance, although that's getting a bit complicated!

6.9 Clip-on Filters

At several points in the text we have talked about restricting receiver bandwidth for signal-to-noise improvement and noise measurements. A useful tool for this work is a small set of clip-on low-pass filters. These can be made on scrap pieces of blob board with flying leads to a couple of small alligator clips (Fig. 6.12a). A few values (1 kHz, 10 kHz, 100 kHz, etc.) will allow you to deal with gain peaking in a transimpedance receiver and estimate the frequency spectrum of your noise, even without a spectrum analyzer.

For spot noise measurements the low-pass is not very useful, but a passive LCR bandpass filter can do a good job (Fig. 6.12b). At DC the capacitor blocks transmission, while at high frequencies the inductor blocks it. Transmission is a maximum at the intermediate series resonant frequency:

$$f_r = \frac{1}{2\pi\sqrt{LC}} \tag{6.1}$$

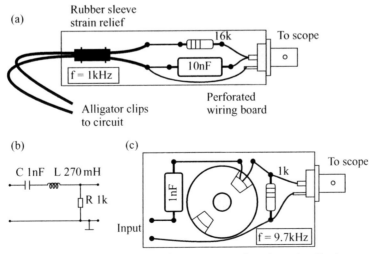

Figure 6.12 Simple, passive filters have many uses in noise estimation in prototype optical detectors. Clip-on RC lowpass and LCR bandpass filters are easy to construct.

the sharpness of the bandpass characteristic is given by the resonance Q factor:

$$Q = \frac{1}{R}\sqrt{\frac{L}{C}} \tag{6.2}$$

If we want to connect this filter to an opamp output then its impedance at resonance should be high enough to avoid overload, say, $1\,\mathrm{k\Omega}$ for a $1\,\mathrm{V}$ output. This means choosing the filter resistance greater than $1\,\mathrm{k\Omega}$. For measurements in the audio range, a few trials of component values show that we need a rather large inductor and small capacitor, and this necessitates using a ferromagnetic core for the inductor. Ferrite pot-cores are about right for this, which are characterized by a value A_L. This is the inductance in nanohenry for a single turn on the core, whereby the inductance goes as the number of turns N^2. A_L is a characteristic of the actual component, not only of the ferrite material, and depends most strongly on the air-gap dimension. Small gapped cores, such as the RM-series from MMG-Neosid have A_L values from 160 to $400\,\mathrm{nH}$. Ungapped RM-series cores in F44 or F9 materials have $A_L = 2000$ to 4300. These are the ones to use for a simple audio filter. For example $N = 250$ turns on an RM10 pot-core with $A_L = 4300$ should give $L = 270\,\mathrm{mH}$. With a $1\,\mathrm{nF}$ capacitor this gives a resonance at $9700\,\mathrm{Hz}$. With $R = 1\,\mathrm{k\Omega}$, $Q \approx 16$. This would be useful to perform noise measurements in the audio band, with the response $3\,\mathrm{dB}$ down at 9420 and $10010\,\mathrm{Hz}$. The RM10 core is about $30\,\mathrm{mm}$ diameter and $20\,\mathrm{mm}$ high and comes with a coil former making winding simple. Again, a scrap of blob board does a good job (Fig. 6.12c). A wide variety of toroidal cores is also available, but these are tedious to wind without elaborate jigging. It is easy to see why we have chosen an ungapped core; the much smaller A_L would necessitate an inconvenient number of wire turns.

6.10   Metal Box for Testing

All optical receivers need very effective electrostatic screening to reduce pickup of line-frequency interference. This means mounting them in a conducting enclosure such as of die-cast aluminum or metal-sprayed plastic. The conducting surface must be connected to the circuitry ground. Small enclosure holes and even the photodiode package window can give problems with high-sensitivity receivers. The same difficulties arise during testing, and so another useful tool is a metal box. This should offer excellent electrical and optical screening, allow quick mounting of circuits under test, and have a range of feed-throughs for power leads and input/output signals. I use a large (300 × 300 × 150 mm) diecast aluminium box. The lid is used as the base, mounted on tall rubber feet or standoffs to make room for a collection of right-angle BNC sockets and power leads. The deep lid should be easily removable but needs to be bonded to the base with a heavy ground lead. Halfway across the lid/base is an aluminum plate wall with a 5 mm hole in it. This is to allow separation of an LED source from the receiver, with just a small hole to couple the two functions. All internal surfaces and the dividing wall should be spray-painted matte black. The lid and box are usually fairly well sealed by the die molding used. If you find that *any* light gets in, glue some angle-aluminum around the edge of the lid/base to shield the join. The use of the lid makes mounting of circuits under test more convenient, and the internal dividing wall can be butchered and replaced to suit the circuits. The good suppression of fluorescent room lights and electrical noise eases measurements of the fundamental noise performance of the receivers. If you have problems with power-supply-borne interference, there is room inside for batteries.

6.11 Butchering LEDs

It is often helpful to remove the molded lens from "water-clear" visible LEDs to allow external optics to be used or just to allow an optical fiber to be placed closer to the emitting chip. The construction of these LED packages is usually a metal header with a depression in it, in which the LED is bonded, used as the substrate contact. A gold wire bond makes the other connection, coupled to the other lead. The form of the wire bond varies considerably between different devices; some are in the form of large loops, while others are almost horizontal from chip to lead. You can easily see the internal construction using a small loupe.

Most of the lens needs to be ground away with a file or coarse grinding paper. You can grip the LED between your fingers or in a pair of pliers, but it is preferable to make a simple jig. One approach is to take a small sheet of plastic or aluminum, 6mm thick and about the size of a PC mouse. Drill a hole to take the LED in the center of one edge. Try to get it tight, so that the LED pushes in, with the chip just showing out the other side. Alternatively, make a screwed clamp to press the LED into an undersized hole. Glue spacers such as two 12-mm washers or scribed and broken pieces of microscope slide glass to the other edge (Fig. 6.13). The idea is to have a three-point mounting, with the LED as one point.

With the LED firmly fixed remove most of the lens with a file, then go to fine waterproof grinding paper (800 to 1000 mesh is good). A piece of float glass makes a flat support. Grind the three points with a circular or figure-of-eight motion until the LED surface looks uniformly matte. It is worth doing this with the LED powered up, just to make sure you haven't gone through the wire-bond! Try to get to within $250\,\mu$m of the bond, checking frequently under the loupe or stereo microscope. Then clean it all off carefully under a running tap with tissue paper or cotton-wool sticks.

Polish it on something finer. A tiny dot of $6\,\mu$m diamond paste on a piece of white bond paper works well, or a creamy slurry of alumina or cerium oxide powder on a cotton cloth, or even toothpaste will work. A minute of movement on the paper will bring the LED surface to a glass-like finish, free from scratches. Just wash off the paste with water or isopropanol.

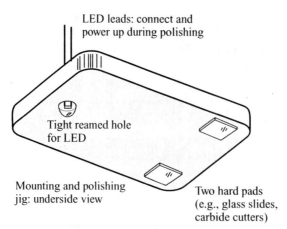

LED leads: connect and
power up during polishing

Tight reamed hole
for LED

Mounting and polishing
jig: underside view

Two hard pads
(e.g., glass slides,
carbide cutters)

Figure 6.13 Removing the lens from an LED is a five-minute job if you have a polishing jig. A "three-point" support gives quick and controllable results. Ideally, power up the LED during polishing to be sure you haven't planed off the wire-bond!

This brief collection of techniques has been designed to quickly get the simple optics lab jobs done, to allow a high flux of didactic experiments. There is little merit in planning a new idea for months, ordering machined parts, and purchasing some micromanipulators, only to find ten minutes after switch-on that the whole idea was half-baked. Yet this is a common occurrence in university and industrial labs. In my opinion most novel ideas should first be tested with whatever bits and pieces you have, on the kitchen table if necessary. With a very modest collection of tools, those first sixty minutes can provide the biggest fraction of what there is to learn, and a number of such experiments provides many occasions to learn.

7

Control of Ambient Light

7.1 Introduction

Apart from some optical fiber communications links, most optical detection systems do not enjoy the luxury of illumination just by the signal of interest; when the source is off, the detector is dark. The great majority are bathed in a background of ambient light from other sources (Fig. 7.1). The signal of interest may even be far weaker than the disturbing light. Clearly we need drastic measures to beat this level of adversary. Mechanical optical screening of either the photodetector, the complete apparatus, or the lab windows with shields and opaque curtains is an effective solution to stop light from entering, but much apparatus under development is operated in an open state. It is very difficult and even dangerous to debug optoelectronics in total darkness, so a compromise has to be reached. In addition, some optical systems such as TV remote controls, bar code readers, light-barriers, free-space communications links, and environmental monitoring systems are required to be open to ambient light.

At the very least we need systems that are unaffected by fairly constant light sources such as overcast skies and by the slow variations caused by moving clouds and people. Suppression of the main 100/120 Hz signals from incandescent and fluorescent lighting is also mandatory. To combat these problems we have emphasized the great benefits of operating our optical systems at a frequency well above DC, and we will assume in this chapter that this has been achieved. Although the signal processing techniques of modulation and synchronous demodulation greatly reduce the degrading effects of such signals, they do have their limitations, so the demodulator should receive as clean a signal as possible. High background light levels will inevitably increase noise and may overload the receivers. Hence we need to control ambient light and suppress its effects. The question is, what characteristics of the desired signal of Fig. 7.1 can be used to separate it from the general ambient light background? There are many approaches, some optical, some electronic; we will look at a few here.

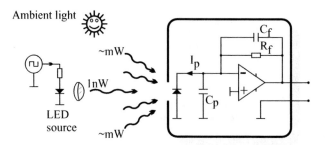

Figure 7.1 Many laboratory experiments are bathed in interfering light, which can sometimes be much more intense than the desired signal. The first defence weapon is modulation.

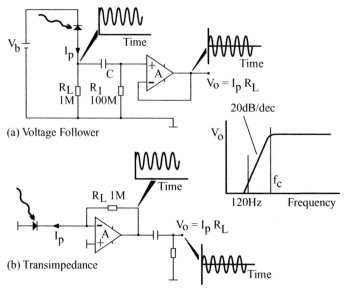

Figure 7.2 AC coupling is not much help to reduce the DC response, as the front end can still overload. Choose f_c high enough to get adequate suppression at the interfering frequency.

7.2 Frequency Domain

7.2.1 AC Coupling: voltage follower

The first characteristic difference between our weak modulated signal and the strong ambient light is of course its modulation frequency; ambient light intensity generally doesn't change rapidly. To help the synchronous detection process we could therefore try to suppress the receiver's DC and low-frequency response, for example by AC coupling the first stage. Where a bias box and voltage follower are used, this is straightforward to arrange (Fig. 7.2). Photocurrent due to the incident light, containing both our modulated signal and low-frequency interference, flows through the load resistor R_L. However, the high-pass filter action of C suppresses the signal seen by the opamp with a low-frequency cutoff at $f_c = 1/2\pi C(R_L + R_1)$. The signal modulation frequency should be $>f_c$. The rate of cutoff below f_c is $-20\,\text{dB/decade}$ of frequency, so with $f_c = 1\,\text{kHz}$, approximately $20\,\text{dB}$ suppression (to 10 percent in voltage) will be obtained at the dominant $100/120\,\text{Hz}$ interference frequency. Slower intensity variations will be suppressed further. If higher suppression is needed, the modulation could perhaps be moved to $10\,\text{kHz}$ or beyond and f_c increased proportionately. Alternatively further RC networks can be cascaded for a greater than $-20\,\text{dB/decade}$ slope. If $100/120\,\text{Hz}$ interference is the main problem, modulation at a frequency several decades above $100\,\text{Hz}$ will greatly ease filtration by AC coupling.

While effective, this circuit can still be overloaded by high-intensity low-frequency illumination. This is because low-frequency photocurrents see the full load resistor R_L, and the voltage on it may increase toward the bias voltage V_b, removing the reverse bias condition. The load resistor must be chosen small enough to avoid this, which might mean that it is not as large as we would like for best S/N. We have also seen that the voltage follower configuration is somewhat restricting for high-speed detection.

7.2.2 AC Coupling: transimpedance amplifier

When a transimpedance amplifier is used, AC coupling cannot be added in series with the photodiode. Without a DC path, the anode will charge positively and forward bias it. Instead RC coupling can be used only after the front-end amplifier (Fig. 7.2b). While this again gives a high-pass response and suppression of DC and low-frequency photocurrents before all later amplifier stages, it does not stop the first amplifier from being overloaded by low-frequency light. Again R_L must be small enough to avoid overload, and the receiver's dynamic range will be limited just as with the bias box or Fig. 7.2a.

7.2.3 Inductive coupling

An apparently attractive alternative to series capacitive coupling is shunt inductive coupling. Figure 7.3 shows voltage follower and transimpedance configurations with inductive loads. At DC the inductor will provide a low-impedance path for photocurrent, giving a low output voltage. At high frequencies the inductor will exhibit a high impedance, allowing the full load resistor sensitivity to be obtained. This is what is wanted. It leads to a high-pass response to photocurrent, with a cutoff frequency f_c occurring when the impedance of the inductor ($2\pi f L$) equals that of the resistor R_L. The difficulty with the technique lies in obtaining suitable inductive components.

For example, for a $1\,\text{kHz}$ cutoff and $R_L = 1\,\text{M}\Omega$, we can calculate $L = R_L/2\pi f_c = 159\,\text{H}$. Obtaining an inductor of such a high value, in a convenient package, whose impedance is not dominated by interwinding capacitance, is very diffi-

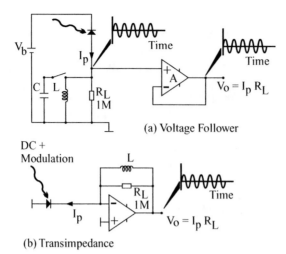

(a) Voltage Follower

(b) Transimpedance

Figure 7.3 Inductive coupling to obtain a high-pass response looks attractive, but the inductor values needed are impractical for low frequencies and high resistance. LCR tuned front ends can be useful in narrow-band systems with high levels of interference and at high frequencies.

cult. A better approach is to add a capacitor across the inductor to form a parallel resonant circuit. The impedance is low at DC and high frequencies because of L and C, respectively, but high on resonance. The load R_L can be adjusted to set the damping constant, which is a compromise between out-of-band suppression and ringing when excited by the modulated signal. The inductor, usually wound on a ferrite core, must be dimensioned to avoid magnetic saturation from the DC photocurrent. Despite the more complicated design, the approach is useful in high-frequency communication systems and in narrow-band, high-interference systems such as 40 kHz modulation frequency handheld remote controls. Performance under intense ambient illumination can be significantly improved by these techniques. Nevertheless, the use of inductive transformer and tuned loads is usually restricted to lower impedances and/or higher frequencies.

7.2.4 Active load/gyrator

Another approach to DC response suppression is shown in the circuits of Fig. 7.4, published by Zetex. These use an active load for the photodiode made up of an RC network and single transistor amplifier. To understand this circuit, imagine a large DC photocurrent I_p flowing through the photodiode and load resistor and then into the transistor base capacitor C_L. As the capacitor charges positively, the transistor will become forward biased, turning it on. This shorts out the load resistor, limiting the output voltage.

On the other hand, high-frequency photocurrents appearing on the transistor base are routed to ground through the capacitor, stopping the transistor

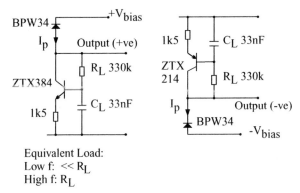

Equivalent Load:
Low f: $\ll R_L$
High f: R_L

Figure 7.4 A nonlinear transistor load can synthesize the operation of an inductor, with excellent suppression of DC and low-frequency light. Low frequency photocurrents forward bias the transistor and short out the load resistor. Circuit reproduced by permission of Zetex Ltd.

from being turned on and allowing the high-value load resistor to be fully utilized. The transition between low- and high-frequency behavior occurs when R_L and C_L have the same impedance. Using the values shown, the low-frequency impedance is about $1\,k\Omega$, rising to $250\,k\Omega$ at $50\,kHz$. The circuit is a simple example of an electronic gyrator, which uses a capacitor and gain stage to synthesize a large value inductor. The output must be taken to a high-impedance buffer such as a voltage follower. Negative output voltages can be achieved by switching transistor polarities, as shown.

With a little manipulation the active load can be used in a transimpedance configuration (Fig. 7.5) to give DC suppression and speed advantages compared with the same load and voltage follower. This network does, however, lead to an offset voltage of about $0.8\,V$ at the amplifier output.

7.2.5 General active feedback

In principle, any network can be used to provide the frequency-dependent feedback around a transimpedance amplifier that is required to suppress the DC response to static ambient light. By exchanging the single transistor of Fig. 7.5 for an opamp we can improve performance and ease the design compromises. This alternative approach has been published by Burr-Brown (1993).

The circuit (Fig. 7.6) is designed for use with the transimpedance configuration of photoreceiver. It consists of a conventional receiver, here drawn with a $100\,M\Omega$ feedback resistor, with additional feedback from the circuitry in the shaded box. This is a essentially a low-pass filter with cutoff frequency $f_c = 1/2\pi R_1 C_1$. As it is in the feedback path, the overall transimpedance has a high-pass response. Well above $f_c = R_L/2\pi R_1 C_1 R_3$ the transimpedance is equal to R_L. Below f_c the response drops off as if there were an inductor across R_L. The extra factor R_L/R_3 comes from the ability of the smaller resistor R_3 to deliver more feedback current than R_L itself, which increases the loop gain.

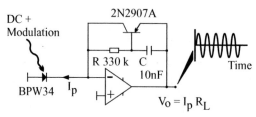

Figure 7.5 The nonlinear circuit of Fig. 7.4 can also be used in transimpedance configuration. DC response is suppressed, although there is a static offset at the opamp output.

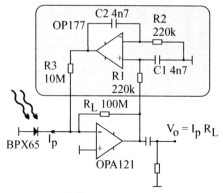

Figure 7.6 The integrator in the active feedback loop also suppresses low frequency response, but maintains signal linearity. Circuit reproduced by permission of Texas Instruments Burr-Brown.

With the values shown, the cutoff frequency is 1539 Hz. This gives a suppression of 120 Hz optical interference of about 25 dB. As mentioned earlier, it is possible to manipulate the frequency response of the feedback path arbitrarily to achieve removal of specific interfering signals. However, maintaining stability and low noise then becomes complicated. The active devices used in Fig. 7.6 are a low-bias current FET opamp for the transimpedance function and a low offset voltage bipolar opamp for the low-frequency feedback. The low-voltage offset is required to avoid a DC voltage on the photodiode. Note that the active feedback does not reduce the shot noise of the DC photocurrent. The full current still flows; it is just split with low frequencies flowing through R_3 and high frequencies through R_L.

Figure 7.7 shows the resulting performance of such an ambient-suppressed receiver. The upper curve is with a simple 1 MΩ load resistor. We can see the rectified sine-wave interference due to 100 Hz fluorescent lighting and the unresolved 10 kHz modulation from the source. The lower trace is the same circuit with the frequency-dependent feedback connected. The modulation amplitude

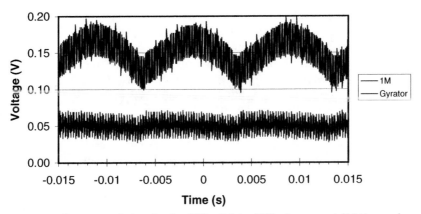

Figure 7.7 Response of the circuit of Fig. 7.6 to 50 Hz fluorescent lighting and a 10 kHz modulation. The 100 Hz rectified sine wave of the upper response is almost suppressed with the active feedback circuit shown in the lower curve.

has not been affected, but the magnitude of the interference has been reduced by a factor 15. Lower frequency offsets due to sunlight are suppressed even further.

7.2.6 Integration

Interference from line-frequency signals (50/60 Hz) and harmonics is of course not limited to optical measurements. All handheld multimeters suffer from similar interference, and in their design a simple form of digital signal processing is often applied. If the instrument's analog-to-digital converter is configured to integrate the input signal for a period of one (or more) whole cycles of the interference (16.7 or 20 ms), then disturbing signals with a zero mean are averaged close to zero. Significant suppression also occurs at all harmonics of the disturbing frequency (limited by the sampling frequency used). High-resolution analog-to-digital converters are often pin-switchable for 50-Hz or 60-Hz integration. While this digital filtering is useful for electrical interference at line frequencies, it does not suppress disturbing DC photocurrents. This can be provided by different signal processing algorithms (App. C).

7.3 Wavelength Domain: Optical Filtering

The above electronic techniques can only be considered partial fixes, when the real issue is to stop interfering signals arriving at the photodiode in the first place. Differences in optical wavelength characteristics can be used for this. When the source is a semiconductor LED, organic polymer LED, laser, or even gas discharge lamp, its optical spectrum is likely to be quite different from that of ambient light. Hence we should consider optical filters designed to maximize throughput of the desired signal and suppress the ambient. There are lots of options. Laser sources are the easiest to filter. Their outputs are usually well defined, and stable to about 1 nm in even the least stable solid state laser diode. Hence we can use filters centered on the signal source wavelength and blocking at wavelengths only a few nanometers away. These bandpass filters can be very effective. Thin-film interference filters formed from multiple layers of deposited alternating high- and low-index transparent material are widely available with center wavelengths from about 220 nm in the UV to 2 μm in the near IR, with bandwidths of 10 to 100 nm. Small pieces of these filters, diamond sawn from larger units, are big enough to cover many photodetectors. However, they do have limitations which affect system performance. The transmission in the passband is typically 50 percent, so half of the desired signal is also thrown away. They are often designed for normal incidence and their transmission changes markedly off axis. They always have other passbands separated from the main transmission region, which require additional filters such as colored-glass types to block them. They can be expensive, and last their suppression is not infinite (typically 10^{-3} to 10^{-4}), especially as the light is reflected by the multilayer stack, not absorbed. Nevertheless their performance with simple bandpass configurations with collimated light can be stunningly effective in isolating well-characterized sources from bright ambient light.

Another useful class of multilayer interference filter is the dichroic filter. These are produced in huge volumes as "hot" or "cold" mirrors for halogen lightbulbs and projector mirrors, reflecting most visible light but allowing infrared to be transmitted and removed from the instrument. They offer an edge response, reflecting either to longer or shorter wavelengths (Fig. 7.8). Devices can be purchased for normal incidence and for 45° incidence, and some types are very economical given the high performance delivered.

Recently another class of interference filter has become popular, the rugate filter. This uses a single film of deposited material with a refractive index profile normal to the filter surface which is continuously graded. This extra degree of freedom compared with a high/low refractive index stack allows the fabrication of filters with almost arbitrary transmission spectra. Filter synthesis software tools have also been developed to make this flexibility practical. One application is to make narrow notch filters for eyesight protection which reflect a design laser wavelength, but otherwise transmit to give a nonfatiguing low-color view of the scene (Fig. 7.9). Multiple notches can be made in the same filter. Even color-correction filters with a complex intensity reflection coefficient spectrum can be fabricated in such a manner.

Although they offer much more limited wavelength filtration functions, colored-glass filters, made from absorbing materials in a glass host, can often provide much higher ultimate suppression than interference filters. They are available primarily with long-wavelength pass responses, although a few bandpass filters exist (Fig. 7.10). There is a particularly useful range of long-wavelength passing glass filters from Schott with designations such as WG with 280 nm to 345 nm wavelength cutoffs, and the GG, OG and RG series with wavelength ranging from the 375 nm (deep blue) to 1000 nm

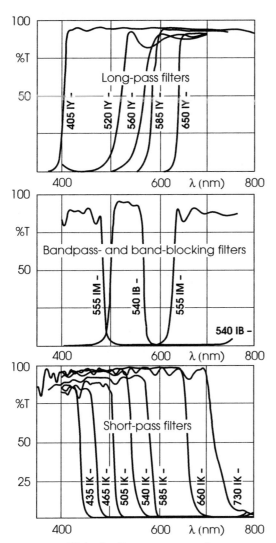

Figure 7.8 Dichroic filter responses. Long wavelength-pass, short-pass and bandpass-types are available. Figure supplied and reproduced by permission of Comar Ltd, Cambridge U.K.

(near IR). These exhibit low passband losses of the order of 8 percent (Fresnel losses at two surface reflections), and stop-band suppression to better than 0.01 percent.

Gelatin, polyester and acrylic filters are also very useful for color correction and control, often called generically "Wratten" filters after Eastman Kodak

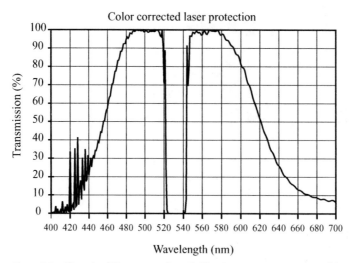

Figure 7.9 Almost arbitrary wavelength filtering responses are possible with the new synthesis tools and manufacturing techniques of "rugate" filters. These use a deposited film of continuously varied refractive index, rather than the Hi-Lo stacks of conventional interference filters. The curve shows a filter for visual use with color-correction and strong suppression of a small band of wavelengths around 532 nm. Figure supplied and reproduced by permission of Advanced Technology Coatings Ltd., Plymouth U.K.

Company's original range. These are cheaper than colored glass filters, but generally do not offer as high suppression. A selection of long-wavelength pass filters is available from about 395 nm (Type "1A") in the blue to 852 nm ("87C") in the near IR. In addition, there are a few bandpass filters in the visible wavelength range. Being plastic, they are easily cut to size for custom applications and quick tests.

Another way to obtain very high suppression of particular wavelengths, especially with a long-pass characteristic, is to use pure, simple liquids. For example ethyl acetate in aqueous solution is transparent in the visible, but absorbs strongly below about 255 nm. Carbon tetrachloride performs similarly. Figure 7.11 shows a dilute solution of acetone ($\approx 50\,\mu l/ml$ in water). The absorption is a smooth peak centered on 270 nm (lower curve), which is useful to absorb the 254 nm mercury line. As the concentration is increased (upper curve), this looks more and more like an edge filter, which shifts to longer wavelength. It remains very transparent in the visible range. The ultimate absorption is much greater than 3AU, which is the measurement limit of the spectrometer used for this spectrum. The convenience of a liquid filter lies mainly in wide-range adjustment of the absorption strength (via concentration), very high absorption at large thicknesses and the ability to form complex shapes. Transition metal ions in aqueous solution are also useful, having strong visible colors which can be used to absorb relatively narrow spectral regions. These metal ions often can

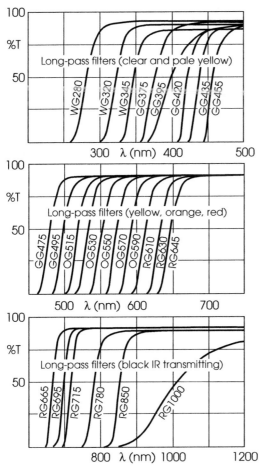

Figure 7.10 Examples of colored glass filter responses. Edge filters are available with modest slopes and very strong blocking of out-of-band wavelengths. These are very useful for many instrument applications. Figure supplied and reproduced by permission of Comar Ltd, Cambridge U.K.

exist in different valencies and oxidation states, giving a range of absorption spectra for a single ion. Several can even be switched between states via the solution pH. As pH can be electrically controlled, so can the optical filter response.

Gases can also be useful in a few situations. Bromine, chlorine, iodine, and sulphur dioxide have strong absorptions in the 300 to 600 nm range, adjustable via their vapor pressure, or by their concentration when dissolved in water or carbon tetrachloride. Even if the absorption spectrum is highly complex, its strength can be tuned using pressure. This approach can be used to make an

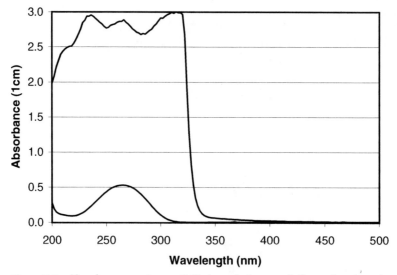

Figure 7.11 Absorbance spectrum of dilute and strong solutions of acetone in water. Many chemicals have useful filter responses, especially edge filters in the UV. Their great advantages are strong blocking, easy absorbance adjustment, and shape flexibility.

absorption modulator which is matched to modulate the received power of a specific absorption. This technique is discussed briefly in Chap. 10.

7.4 Polarizers

Polarization coding can be applied both in passive and active approaches for ambient light suppression. In the simplest case the signal is polarized before transmission, and a matched polarizer is used at the receiver. If the disturbing light is unpolarized, its intensity will be reduced by 50 percent; not great, but even this may be useful. The suppression can be much greater if the disturbing signal is itself polarized. For example, Fig. 7.12 shows a turbidimeter designed to see below the surface of an outdoor flowing water stream. The signal source is a linearly polarized diode laser, arranged with its polarization in the plane of incidence. Light that refracts into the water is multiply scattered, which greatly reduces the degree of polarization. Hence the returned scattered light is almost depolarized, and 50 percent of it passes through a polarizer to reach the detector.

When used as designed in the open air, these sensitive receivers can easily be overloaded by intense sunlight reflecting off the water surface. Even with the best orientation of the turbidimeter to avoid direct sunlight reflection, the diffuse sky illumination seriously limits performance, even in the United Kingdom! A solution is to operate the sensor close to Brewster's angle (Fig.

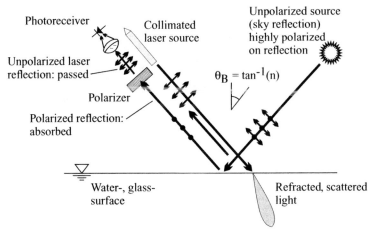

Figure 7.12 Polarizers can be useful to suppress highly polarized interfering sources, for example from daylight reflections off surfaces close to Brewster's angle. This example is from a laser-based water turbidimeter for outdoor use.

7.12), given by $\theta_B = \tan^{-1} n$, where n is the refractive index of the water. For $n = 1.33$ θ_B is 53°. At this angle the reflected light is strongly polarized perpendicular to the plane of incidence. A polarizer oriented in the plane of incidence as shown can achieve suppression to better than 0.1 percent (plastic polarizers) or even 0.01 percent (crystal polarisers), as long as the acceptance angle of the receiver is less than a few degrees. Good suppression using plastic polarizers is, however, achieved at the expense of higher absorption losses. For incident and refracted angles θ and ψ and incident and refracted medium indices n_1, n_3 we have:

$$n_1 \sin \theta = n_3 \sin \psi \text{ (Snell's law)} \tag{7.1}$$

The reflected amplitudes are given by (Longhurst 1968):

$$A2s = \frac{-\sin(\theta - \psi)}{\sin(\theta + \psi)} \tag{7.2}$$

$$A2p = \frac{\tan(\theta - \psi)}{\tan(\theta + \psi)} \tag{7.3}$$

And the transmitted amplitudes are:

$$A3s = \frac{2 \sin \psi \cos \theta}{\sin(\theta + \psi)} \tag{7.4}$$

$$A3p = \frac{2 \sin \psi \cos \theta}{\sin(\theta + \psi) \cos(\theta - \psi)} \tag{7.5}$$

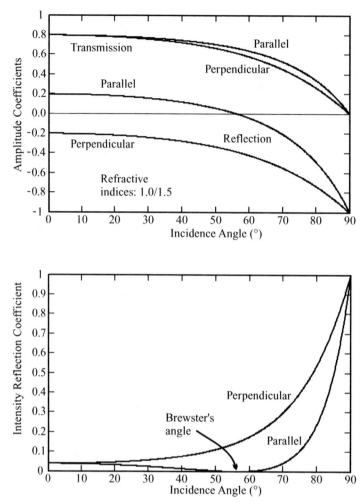

Figure 7.13 The amplitude and power reflection coefficients are given by the Fresnel equations. At Brewster's angle the reflection coefficient for parallel-polarized light vanishes, giving a highly polarized reflection.

The intensities are obtained by squaring these amplitudes. The s and p correspond to perpendicular and parallel electric vector polarizations respectively, 2 and 3 to the reflected and refracted amplitudes. Figure 7.13 shows the amplitude and intensity reflection coefficients of glass for a range of angles.

7.5 Field of View Restriction, Absorption and Mechanical Approaches

Next we come to mechanical screening techniques, at once trivial and complex. This is primarily a question of knowing where interfering light sources are, and tracing the possible rays to your instrument and your detector. For a free-space-optical system where direct sunlight and diffuse sky radiation are disturbing, placing simple opaque "snouts" on receiver input-windows can be very effective, either with or without additional collimating optics (Fig. 7.14). Their effectiveness depends on the acceptance solid angle, and hence is improved by increasing the aspect ratio, length to width, of the tube. Small angles give better protection, but can make the required pointing accuracy intolerable. Some designs have used thin metal "potato-cutter" structures to give simultaneously a large aperture and small acceptance angle in a compact structure. A low-reflectance coating or internal apertures inside the snout will reduce light transmission by multiple scatter. Painting the inside of the tunnels black is surprisingly ineffective, mainly because even matte black paints are highly reflective at grazing incidence. For outdoor applications snouts also offer some protection against rain, although drain-holes should then be provided. In dusty environments the use of long aspect ratio tunnels can protect against dust build-up. In extreme environments a flow of clean air or water pumped through and out of the snout will be needed. Even electrostatic deposition electrodes may be called for.

In the design of high sensitivity dark-field measurements, for example in a compact turbidimeter (Fig. 7.15), absorbing, rather than simply reflecting the unwanted light is a big advantage. This application requires great care if good performance is to be obtained. You will want to use a high powered source for best sensitivity, although the ultimate performance is likely to be defined by the fraction of stray light, rather than the sensitivity of the photoreceiver. Try to avoid any light impinging of the cuvette corners, using thin, accurately sized apertures; they are points of strong scattering. Any light incident on the silica walls out of the beam-paths should be absorbed in coatings. These are very helpful, but difficult to maintain. The only way to minimize the power reflected

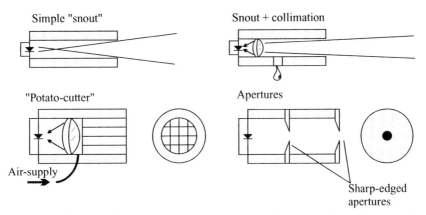

Figure 7.14 Even simple snouts and apertures can greatly improve the performance of receivers subjected to intense ambient light.

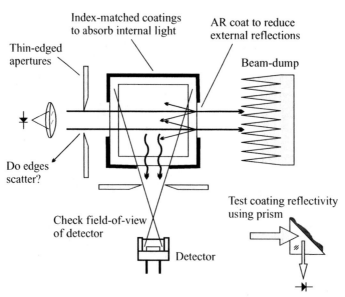

Figure 7.15 Sensitive light scatter and fluorescence measurements require obsessive attention to avoiding or absorbing all stray light. Internal reflections in a small cuvette can be reduced by coating the external surfaces with an approximately index-matched absorbing material. One way to evaluate these is to measure reflection coefficient in a 45° prism.

at the silica/coating interface is to match the complex refractive indices of the two materials. The power reflectance formulas of Eqs. (7.1) to (7.5) are valid even when the refractive indicies are complex. However, even if you could match both real and imaginary parts of n, it would just put off the problem to the next interface. The best approach is to find a substance which is a good match of the (real) refractive index, which is also moderately absorbing (not a metal). For example, a perfect match of real refractive index of a 1-mm-thick coating with an absorption length of 100μm should work quite well. The 100μm figure implies that the imaginary part of the refractive index, that is the absorption, is very small, and hence will not contribute to a large impedance mismatch between silica and coating. Spray-paints and acrylic coatings from art shops are as reflective as glass when viewed from the air, but can be extremely effective when viewed from inside the silica material; a few experiments will be needed to choose the best materials. It is easier to screen available paints and gels by applying them to the hypotenuse of a 45/45/90° silica prism rather than a flat substrate. In the latter case it is too difficult to separate the front- and back-surface reflections unless the flat is very thick. Some weakly absorbing acrylic paints can achieve better than 0.1 percent power reflectivity in the visible when coated onto silica and glass.

To absorb the straight-through free-space beam rough paints containing black flocks are available and useful, but can be polished and degraded by inappropriate handling. Black velvet is available, even self-adhesive, which is generally even better. Some of the thick carbon-loaded plastic foams used for electrostatic protection and packaging of integrated circuits can have very low reflectivity, of the order of 0.1 percent in air, but 10 mm or more may be necessary due to the tenuous porous structure.

Blackening treatments for metal are available and are usually performed electrochemically. If the original metal is highly polished, blackening to reduce reflections is of limited use (but they are attractive). Some materials can be effectively textured to give an interface of mechanically-graded refractive index. For example, silicon can be treated by ion-beam etching to give a dendritic structure with a depth much greater than the wavelength of visible light. Such surfaces can have an almost magically invisible appearance due to their low reflectivity. Stacks of new razor blades (Fig. 7.16a) can function in a similar way, with the sharp edges limiting scattered energy, and multiple reflections between blades gradually absorbing penetrating light. This implies that absorbing, highly polished surfaces may perform better than roughened ones, as they limit the change in angle after many reflections. The blades can also absorb considerable energy without damage.

Achieving absorption efficiency is often easier if you have lots of space. One of the best absorbers is a large hollow cavity with a small entrance hole (Fig. 7.16b). The cavity should be diffusely reflecting and blackened using any of the above techniques, although the main effect in use is the distribution of the input light over a large internal area, so that only a small fraction escapes through the hole. Even if the cavity were white and diffusely reflecting, a 1-mm-diameter hole in a 1000 cm^3 cavity would only allow about 10^{-5} of the input light out. Blackening will get you another order of magnitude or more. Infrared black-body sources are often fabricated in this way, with a small hole leading to a large, heated internal cavity.

(a) Razor blade Stack   (b) Black-body

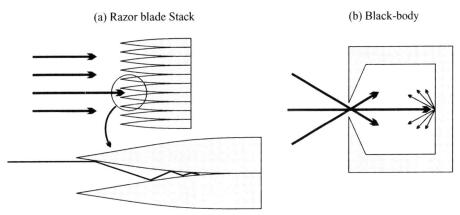

Figure 7.16 Stacks of new razor blades can form an effective, high power beam dump via multiple reflections and absorptions. A small aperture in a large, diffusely reflecting cavity also works well.

It is important that the apertures is sharp, the idea being to present only a small area which scatters light. These can be fashioned very effectively from thin metal foils. Don't just try to use a drill press with the foil on a soft support. Holes can be drilled cleanly by first clamping the foil tightly between two thick scrap plates, ideally of a similar material. Use a sharp drill bit, lubricate with oil-emulsion or kerosene, and run the drill-press at the correct *linear-velocity* for the bit. Feed-rate is also important, and a few attempts to get it right will be needed. If it is right, edge damage can be less than the foil thickness. This even works well with the thinnest polymer foils such as Saran Wrap, which are impossible to drill directly. For the thinnest metal foils ($<25\,\mu$m) photolithography and one- or two-sided etching is a good alternative approach.

7.6   Coherent Detection and Localization

The difficulty of making sensitive optical measurements in a confined space occurs frequently. Let's imagine we want to detect the light scattered from a small dust particle drifting around the room (Fig. 7.17). We use a collimated laser beam, but if the scattering efficiency is only 1 in 10^6, we get just 10 nW back from our 10 mW laser. The majority of the light passes by to illuminate the apparatus walls, contributing to scatter which could be far more intense than the signal from the particle. The whole experimental environment is illuminated by an interfering source. We could of course modulate the laser light and synchronously detect, and this would be effective in separating it from room light. However, all the disturbing wall-scatter is equally modulated, and will not be filtered out by the lock-in process. This is an opportunity for coherent detection.

Figure 7.17*b* shows an option. The detector signal is made up of the weak particle-scatter and a reference beam obtained by splitting a fraction of a percent from the main beam. We will assume that the coherence length of the laser source is much longer than the difference in path-lengths beam-splitter/particle/detector and beam-splitter/detector. Then the two signals will "interfere," meaning that we must calculate the detected intensity by first summing the light amplitudes, not powers. For the two complex amplitudes we write:

$$\vec{A} = Ae^{j\phi 1} \tag{7.6}$$

$$\vec{B} = Be^{j\phi 2} \tag{7.7}$$

where A, B are the real field amplitudes and ϕ_1, ϕ_2 are the phases of the two waves. The total detected amplitude is then:

$$\vec{E} = \vec{A} + \vec{B} \tag{7.8}$$

and the intensity I is proportional to $\vec{E} \cdot \vec{E}^*$, where the $*$ denotes the complex conjugate. We obtain:

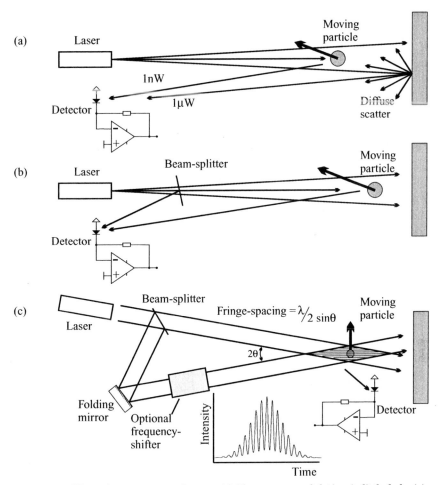

Figure 7.17 Where intense scatter is unavoidable, source modulation is little help (*a*). Coherent detection effectively modulates the received signal depending on the particle's velocity (*b*). (*c*) An alternative configuration which allows more flexible choice of the modulation frequency.

$$I = A^2 + B^2 + 2AB \cos \Delta \qquad (7.9)$$

where $\Delta = \phi_2 - \phi_1$ is the phase difference. As the particle moves toward the beam-splitter the scattered signal phase will change with time, and with it the detected intensity. In this case we do not need to modulate the laser to obtain an intensity modulation. If the particle is moving with a constant v m/s, then we will detect a sinusoidal beat. With the almost collinear geometry shown, the beat signal will have a frequency of $2v/\lambda$ (Hz). For a $0.633\,\mu$m wavelength helium neon laser and a velocity of 1 m/s this is about 3.2 MHz. Hence the particle has provided its own unique modulation signal, coded to represent its velocity. All particles moving at 1 m/s contribute to the 3.2 MHz signal. All 1.1 m/s particles give a 3.5 MHz signal and so on. Most importantly, the stationary wall does not give appreciable modulation, and so its scattered light signal can be electronically separated from that of the moving particles.

The backscatter geometry of Fig. 7.17b is restricting for some applications. An alternative is to form an interference pattern in the intersection of two beams. As a particle moves through the pattern, it scatters a chirp of light with a characteristic frequency given by the number of fringes-per-second intercepted. In this case the detector can be placed almost anywhere where it can pick up adequate signal. The ability to tailor the fringe spacing through adjustment of the intersection angle allows it to be optimized for particles of a particular size. The chirp frequency gives the component of particle velocity normal to the fringes, while the visibility of the pattern depends on of the particle size.

Figure 7.17c shows an unbalanced interferometer, in which the two arms have very different lengths. This is not a problem for a source of greater coherence length, and high contrast fringes will be formed throughout the intersection volume. On the other hand, if a very short coherence length source such as an LED were used, no fringes would be seen in the diagram. This characteristic can be used to *localize* the fringe system wherever desired, for example across a section of a pipe carrying fluid (Fig. 7.18). The symmetry of the beam-splitter and folding mirrors defines a place where the path-lengths of the two interfering beams are equalized. The fringes will only be formed in a narrow sheet which is conjugate with the beam-splitter. If the coherence length is only 25 μm, then a 25 μm thick slice will show good contrast fringes, and hence modulation for moving scattering particles. This can be used to localize the measurement to a single pipe section. Particles moving outside the sensitive region will still scatter light, but this will just contribute to a background DC signal. This gives its shot noise, but can be largely filtered out by detecting at the high self-modulation frequency. Note that this "homodyne" detection

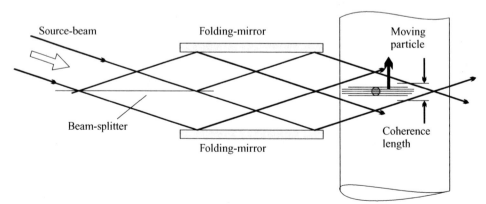

Figure 7.18 Low coherence-length sources can be used to localize the interference pattern in a confined region.

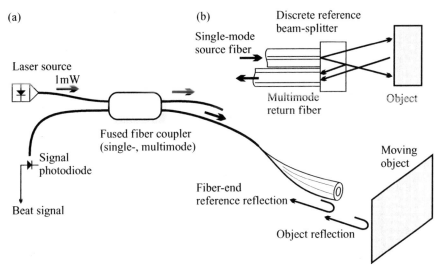

(a)

Laser source
1mW

Signal
photodiode

Beat signal

Fused fiber coupler
(single-, multimode)

Fiber-end
reference reflection

Object reflection

(b)   Discrete reference
            beam-splitter
Single-mode
source fiber

Multimode
return fiber

Object

Moving
object

Figure 7.19 By forming the interference with a reference reflection at the distal fiber end, the delivery fiber length is unimportant; even multimode fibers work well. However, speckle is still disturbing, so a singlemode fiber and tiny discrete beamsplitter can perform even better.

cannot distinguish between particles moving up or down the page in Fig. 7.18; they both give the same beat frequency. If this is a limitation, it can be removed by frequency-shifting one arm of the interferometer using an acoustooptic modulator as in Fig. 7.17*c*. This has the effect of causing the fringe pattern to move in one direction. Then particles moving with the pattern show a lower frequency modulation; those moving against the fringe pattern motion show higher frequency scattered light modulation. The differences can then be separated out using electronic filtration techniques. This is called a *heterodyne* detection system.

Simple homodyne interferometers lend themselves ideally to construction in fiber-form. Figure 7.19*a* shows a fiber-coupler based interferometer, in which the reference signal comes from the Fresnel reflection at the fiber end. This amplitude is added to that from the object to give an interference beat which can be detected via the beam-splitter. This geometry is very useful as it refers the interferogram to a plane close to the moving object. However, if a multimode fiber system is used there will still be signal fluctuations due to speckle in the fiber. As the emitted light field is made from a large number of effective modes, each of which may have a different optical phase, the detected intensity depends on mechanical motion of the up-lead fiber. This can be alleviated by using a single-mode fiber for the source fiber (Fig. 7.19*b*), an external beam-splitter and a multimode receiving fiber. This leads to a much more stable system in which vibration and movement in the fibers do not lead directly to speckle-noise intensity fluctuations. Some intensity modulation caused by polarization state variation remains, but this is much smaller. The simplicity of such fiber-interferometers means that they can often be integrated into complex machinery such as high-speed actuators. If the object's direction is known, frequency-shifting is not required. The collinear back-scatter geometry means that calibration is not necessary, as one detected fringe means a motion of $\lambda/2$ for the source in use.

The control and suppression of unwanted light can often be as difficult, and as expensive, as detection of the wanted light signal. Success often appears to be down to a random series of trials and errors. Nevertheless, a methodical approach of understanding the origin and eventual absorption of every possible light source, combined with a bit of pencil-and-paper ray-tracing, will usually get you to that success more quickly.

8

Stability and Tempco Issues

8.1 Introduction

Many optical measurement systems suffer from a lack of optical signal energy. Either the source is weak, or very distant, or must be detected in a short time, or suffers from great attenuation on its path to the photodetector. In all these cases, the main design effort is directed at achieving adequate receiver *sensitivity*, where optical and electronic noise and interference are the adversaries. This has been the primary focus of the chapters up to now. Another class of measurement has plenty of light, such that enormously high, shot-noise limited signal-to-noise ratios are possible in principle, and the goal is to measure the optical power associated with the signal with exacting precision over the life of an instrument. Here the main issue is a lack of *stability*.

8.2 Chemical Analytical Photometry— Design Example

Figure 8.1 shows an example of such a measurement, namely of the transmission or attenuation of a liquid sample in a transparent silica cuvette or capillary. This is typical of many applications in analytical chemistry, where changes in attenuation can be related to the (low) concentrations of an absorbing chemical, either directly using their own optical absorptions, or via a colorimetric indicating reagent. Let's assume that the chemical we wish to measure absorbs at 430 nm in the visible blue range. The solvent is water and high power LEDs are available emitting near to this peak absorption wavelength. Light from the LED is collimated and couples through the aqueous sample-filled cuvette to a well-designed photoreceiver. The cuvette outer walls are antireflection coated to give almost zero Fresnel reflection losses, and index mismatches at the silica/water interface (≈ 0.14) give only small intensity reductions (≈ 0.25 percent) at each interface.

The water exhibits an absorption of less than $0.01\,\text{cm}^{-1}$ at this wavelength,

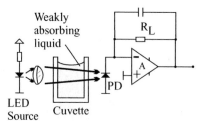

Figure 8.1 Measurement of small, slow changes in transmission represents one of the most difficult measurements.

so that 1 mW of the source light reaches the photodiode and receiver. At this wavelength the photodiode's responsivity is only 0.15, but nevertheless we have 150 μA of photocurrent. A resistor of 33 k delivers almost 5 V output from the transimpedance amplifier, and the design is shot-noise limited. The LED is modulated at 10 kHz, well away from interfering line-voltage harmonics and the low-frequency $1/f$ noise increase. With 150 μA photocurrent the shot-noise limited precision ($\Delta I/I$) should be about 10^{-7} in a 1 Hz bandwidth.

It is instructive to build such a simple system, connect it to a data-logger, and run it over a period of a few hours. Initially the performance may be awful, as the tiny battery you chose to run the LED dies after a few minutes, the double-sided adhesive tape holding the photodiode falls off, and the cuvette walls become covered in bubbles out-gassing from the liquid. After these oversights have been remedied, in any but the stablest of environments, the detected signal can still be expected to vary by much more than 10^{-7}, for example by several percent, primarily due to changes in component parameters with temperature variations.

8.2.1 Load and bias resistors

The photoreceiver's detection gain is largely unaffected by changes in opamp characteristics. As long as AC measurements are made at a frequency where closed loop gain is still high, the open loop gain and the amplifier offset voltages play only a small role. The bulk of the precision of the circuitry rests with the feedback components. The most obvious source of error is therefore the transimpedance R_L. Output voltage is linearly proportional to it, and all resistors change their resistance with temperature. Described by the relative temperature coefficient of resistance, or "tempco" ($1/R \; dR/dT$), and conveniently given in units of ppm/°C (10^{-6}/°C), this parameter depends on the fabrication process. Some types of low cost carbon resistors exhibit a tempco of ±250 ppm/°C or worse. Hence for a 50°C temperature fluctuation in the component, a combination of ambient temperature variation, enclosure heating due to electronic dissipation, and self-heating in the resistor, we can expect a sensitivity variation of ±1.25 percent over the temperature range. The LED bias resistor suffers similar variation, and its self-heating is likely to be greater due to the higher currents flowing.

For very little extra cost, we could use high stability metal film resistors with a tempco of ±50 ppm/°C, reducing the gain variation to ±0.25 percent total. However, this assumes that resistors are available. While the 10 Ω to 1 MΩ resistors used for LED bias and general operational circuitry will surely be available, the 10 MΩ to 10 GΩ needed for a sensitive transimpedance amplifier may be harder to obtain. These high value resistors often show much greater temperature variations. However, with the high intensity we don't have that problem here. If even greater stability is required for analog circuitry, 10 ppm/°C metal film resistors are widely available, and specialist wire-wound and other resistors can be obtained to better than 1 ppm/°C, at a price greater than most operational amplifiers.

8.2.2 Capacitors, lumped and parasitic

The analysis of Chaps. 2 and 3 showed that a photoreceiver circuit's noise and signal output characteristics as a function of frequency depend not only on the transimpedance resistive load, but also on other deliberately introduced components, and on circuit strays such as the parasitic capacitance of the load resistor. We can use the same considerations to calculate output stability with temperature variations. For example, the temperature coefficient of any feedback capacitor used to limit receiver bandwidth or provide stability will affect the signal gain, unless the operating frequency is well below the characteristic break frequencies. Even if no additional capacitance is used in parallel with the feedback resistor, the remaining stray capacitances of the resistor and circuit wiring may dominate temperature stability performance. These parameters are unlikely to be known, well-characterised, or even repeatable from one fabrication to another, so this is probably a situation to be avoided. Although these components are generally neglected in considering gain stability, when working at the limits of a receiver's bandwidth, even if this is only 1 kHz, they can become dominant factors. Hence if everything has been done right, and the tempco is still worse than expected, these capacitive components should be suspected.

8.2.3 Photodiode temperature coefficient

Next we have the photodiode itself. Many photodiodes exhibit a photocurrent tempco $(1/I_p \, dI_p/dT)$ near to +0.25 percent/°C. For example, data-sheets for the well-known BPX65 photodiode (1-mm^2 area, silicon) give a change in photocurrent with temperature of +0.2 percent/°C. This equates to a 10 percent sensitivity change over the above 50°C temperature range, a serious uncertainty in many instrumentation applications. Several years ago Hamamatsu introduced a range of photodiodes with low tempcos of ±100 ppm/°C and less. The tempco is wavelength-dependent, even changing sign. Nevertheless, these have greatly improved the precision of many optical measurements.

8.2.4 LED temperature coefficient

When low tempco photodiodes are used, the real problem is usually the LED that is used as the source (Fig. 8.1). These have variable, and often unspecified tempcos of the order of –0.1 to 1 percent/°C, when driven by a constant current.

8.3 Temperature Compensation

If the intrinsic tempco of the optoelectronic components is still too high, temperature compensation may be attempted. Figure 8.2 shows how thermistors can be used with a photodiode or source LED to improve stability. The best choice of values, ability to interchange devices without recalibration, and variety of packages is available with negative temperature coefficient (NTC) thermistors. At room temperature the resistance of common devices varies by about –3.5 percent/°C, large enough to compensate any LED or photodiode.

The temperature sensitivity of thermistors results from the temperature dependence of majority carrier concentrations (see Sze 1969). This leads to a resistance R for semiconductors that varies as:

$$R = R_o e^{[B(1/T - 1/T_o)]} \tag{8.1}$$

where R_o is the resistance at the reference temperature T_o (usually 25°C = 298 K), and B is a characteristic constant (units K). The temperature coefficient at any temperature T is then given by:

$$\alpha = \frac{1}{R}\frac{dR}{dT} = \frac{-B}{T^2} \, °\text{C}^{-1} \tag{8.2}$$

Temperatures are in K. The constants R_o and B are published in the device's data sheets. For example, the common GM102 thermistor has a nominal resistance of 1 k at 25°C and a B value of 2910 K, from which we can calculate a temperature coefficient α of about –3.5 percent/°C at room temperature (Fig. 8.3a).

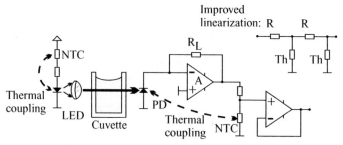

Figure 8.2 Effects of the high temperature coefficients of LED output and photodiode responsivity can be reduced using thermistors closely coupled to them. Ladder networks can improve linearity.

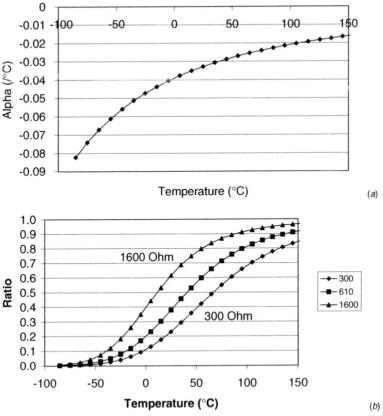

Figure 8.3 (*a*) At room temperature the tempco of many thermistors α is about –3.5 percent/°C. (*b*) The most linear region of a resistor/thermistor potential divider can be shifted though choice of the load resistor.

In use the thermistor can make up one arm of a voltage divider powered by the signal. By choice of the division ratio the region of best linearity can be shifted in temperature. Analog Devices "Transducer Interfacing Handbook" (1981) suggests a value for the series resistor of 61 percent of the thermistor nominal resistance for best linearity. Figure 8.3*b* plots the division ratio with three values for this thermistor with the series resistor connected to ground. In the 0 to 50°C range the 61 percent value does indeed look about right. It is possible to further linearize this characteristic through a combination of series and parallel resistances, at the cost of reduced temperature coefficient, or by using two or more different thermistors in a resistor ladder network. Good linearity makes compensation for the photodiode characteristic easier.

Tiny glass-encapsulated or polymer-coated thermistors are available which can be thermally bonded to the photodiode or LED package. Mechanical clamping, mounting in a common high thermal-conductivity block, and/or bonding with thermally conductive adhesives can all be effective in delivering a temperature value, as measured by the thermistor, which is representative of that of the optoelectronic component. This is especially the case with photodiodes with metal or ceramic packages with good thermal-conductivity.

In the case of lower-cost, plastic-encapsulated LEDs, bonding to the thermistor gives only limited temperature tracking. These LEDs suffer from larger temperature variations due to their significant ($\approx$50 mW) internal dissipation and higher chip-ambient thermal resistance. The coupling can be improved by bonding the thermistor to the LED lead which holds the semiconductor chip (Fig. 8.4), directly outside the polymer encapsulation. The other lead is less useful, as it is thermally isolated by the fine topside wire-bond. If in addition the leads can be cropped short and connected to the PCB with much finer lead wires, isolation from the environment can be improved. Machining away the underside of the LED polymer to position the thermistor in contact with the chip support metalwork performs even better, but is tricky to do. Surface-mount LEDs can also be well coupled due to their very small dimensions.

Of course, thermistors are not the only temperature sensors which can be used for compensation. In principle, all the other available sensor types could be applied, compensating either via analog circuitry as in Fig. 8.2, or by temperature measurement, and compensation in software. Table 8.1 shows a few of the different semiconductor types available, providing either analog voltage or current outputs, or a digital signal most easily read by a microprocessor. However, despite the electronic convenience of the semiconductor ICs, and especially the digital devices, it is difficult to beat the small size and hence mechanical convenience of some of the available bead thermistors.

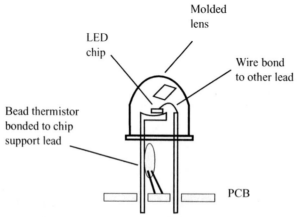

Figure 8.4 Thermal coupling to the chip of a plastic encapsulated LED is most efficient through the substrate lead.

TABLE 8.1 Temperature Sensors

LM19	Analog, $(-3.88 \times 10^{-6} \times T^2) +$
	$(-1.15 \times 10^{-2} \times T) + 1.8639\,V$
LM34	Analog, $10\,mV/°F$
LM35	Analog, $10\,mV/°C$
LM135	Analog, $10\,mV/K$
AD590	Analog, $1\,\mu A/K$
DS1620	Digital, 9-bit coded serial
LJK SMT160-30	Digital, $\approx 250\,Hz$, duty-cycle coded
Silicon transistor V_{BE},	Analog $\approx -2.2\,mV/°C$ or
collector connected to base	-0.04 percent/°C

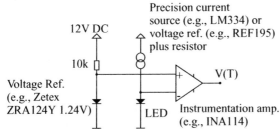

Figure 8.5 The terminal voltage of the source LED itself provides an effective local temperature measure. The LED current should be stabilized to high precision with a current source or voltage reference and low tempco resistor.

Instead of trying to place a discrete temperature sensor as close as possible to the LED chip, an obvious alternative approach is to use the LED junction voltage itself as the temperature sensor (Fig. 8.5). Operated at a constant current, the LED terminal voltage makes a superb internal temperature sensor. Where the LED is current-modulated for use with synchronous detection, as it will almost always be, the terminal voltage can be simply low-pass filtered to obtain a high-resolution temperature-dependent average voltage (Fig. 8.6). Alternatively the AC signal could itself be synchronously detected, reducing low-frequency interference and perhaps improving the signal to noise as with the optical receivers.

Once we have a high-precision, reliable measurement of temperature near or even in the optoelectronic device, there are many techniques to compensate for temperature variations. Figure 8.2 showed a simple open-loop analog technique. Where microprocessor control is available, compensation is more flexible and convenient in digital circuitry. If the long-term drift of LED characteristics means that the power/temperature relationship is not constant, the microprocessor can be reprogrammed with an updated look-up table.

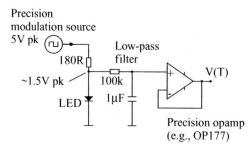

Figure 8.6 Temperature estimation of a modulated LED is also possible by smoothing (or synchronously detecting) the terminal voltage variations.

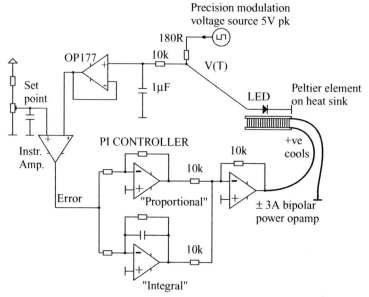

Figure 8.7 Precision temperature control of an LED source can be effected using a PI-controller and Peltier element heat pump. Temperature fluctuations of 1 mK are relatively easy, 10 μK possible with care.

8.4 Temperature Regulation

8.4.1 Simple circuitry

Instead of providing temperature compensation, we can use the brute-force approach of temperature regulation. All of the above temperature sensors are suitable, but for stabilization of the LED temperature the terminal voltage technique appears to be superior. Conventional analog or digital feedback controllers can be used, with heating-only to a regulated temperature above the maximum operating temperature, or heating and cooling using a Peltier element heat pump (Fig. 8.7). Peltier elements allow stabilization near to room temperature, or even below it for photodetector noise reduction. Convenient control and driver modules for Peltier elements are available, driven by the market for controlled telecommunication laser sources, or you can design your own. When stabilizing at reduced temperatures condensation of water onto the source/detector must be avoided by direct bonding or sealing in of a dry atmosphere.

8.4.2 Self-regulating transistors

If heating and temperature stabilization are to be applied locally to a small packaged optoelectronic device such as a photodiode or LED, the physical sizes of conventional sensors and heaters can become inconvenient. Combining these two functions is a useful trick. An elegant technique using a transistor both as sensor element and heater has been published by Woodward (1998). Figure 8.8 is developed from that article. A package of 4 CMOS gates is used to measure the base-emitter voltage of the power transistor, and then heat it by applying short current pulses. By dispensing with a transistor load resistor and relying on its current gain to limit dissipation, the circuit size is minimized and all heating energy goes into the transistor. By using a high current transistor for Q1 in a small package (e.g., Zetex ZTX451 1A in the E-Line package) or a surface-mount device, very compact and low-cost point-of-use temperature controls can be made.

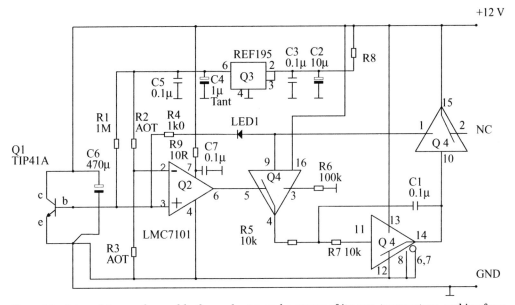

Figure 8.8 A transistor can be used both as a heater and a sensor of its own temperature, making for a very compact arrangement. Circuit after an elegant design by Woodward (1998). Reproduction by permission of the author.

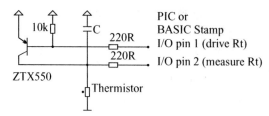

Figure 8.9 A thermistor can also be used as heater and sensor, for example using the RC time-constant measuring routines built in to certain PIC and Parallax Basic Stamp micro-controllers.

Similar techniques can be applied to thermistors, for example using a PIC micro-controller or "BASIC Stamp" (Fig. 8.9). Some of these have built-in functions to determine the value of a connected resistor by timing the discharge of a capacitor. With a transistor switch, power could also be applied. Although dissipation of the smallest devices is limited to a few milliwatts, this is sufficient to regulate the raised temperature of a small packaged device, as long as heat transfer with the environment is minimized with thermal insulation. As mentioned above, their small size makes such an approach even easier to integrate than a transistor heater. It may also be possible to measure and heat a single LED using terminal voltage and current pulsing.

8.5 Optical Referencing

Temperature regulation and compensation are convenient in some optical measurement systems. It is relatively simple to obtain 0.1 K measurement and stability in an on-line industrial instrument, and even few-microkelvin regulation stability in a laboratory environment. See for instance the article by Barone et al. (1995). However, these precisions are still a long way from that required to reach the accuracy and resolution afforded by the shot-noise limit. It is also quite complex to compensate individually for the different temperature coefficients of separate optoelectronic devices, such as a set of assorted-wavelength LEDs, photodiodes, and electronic components. For this, overall temperature regulation seems to be the only solution. Another promising avenue, which attempts to subsume all the errors of a source/detector chain, is optical referencing.

8.5.1 Light taps: division of power

Figure 8.10 shows the schematic liquid optical absorption measurement system, modified for source power monitoring. Just before the cuvette a beam-splitter taps off a fraction of the incident light to be detected at a reference photodetector PD2. If the LED light output varies, the detected signals at each photodiode can be expected to vary accurately in proportion.

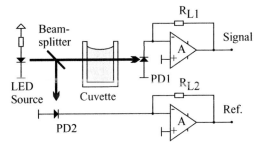

Figure 8.10 Optical referencing can reduce the requirements on LED output stability and greatly improve overall transmission measurement stability. However, the photodiodes must track with temperature, and "asymmetric" window contamination must be kept to a minimum.

There are at least two ways in which the signals can be used. In the most straightforward fashion, both signal and reference are detected in separate amplifier channels, and the signal is then normalized by dividing by the reference signal. Analog dividers such as the Burr-Brown MPY100 and MPY534 were historically used for this function, although their accuracy is limited to approximately 1 percent. These days it is probably simpler, although not necessarily equivalent, to capture both values via analog-to-digital converters (ADC) using a microcontroller or laboratory computer, for division in software. In this case the division is unlikely to add any errors, unless limited by inadequate wordlength integer arithmetic. This brute-force approach requires ADCs for both detected intensities. It is important to sample both channels simultaneously, if the best normalization accuracy is desired. Any sampling delay will be translated via intensity drift into differential errors. If the only power variations are due to slow LED temperature fluctuations, achieving adequate simultaneity should be easy. Higher speed fluctuations can also be reduced, but care in the timing is necessary. We will discuss this further in a later paragraph in this chapter.

Separate measurement and normalization is more elegant than temperature regulation, as it is not necessary to rely on long-term stability of the source temperature coefficient. Further, precision current control of the LED is not needed, which can save another expensive active element. A small modification is to use the reference signal to control the source power output via feedback (Fig. 8.11). If this is performed in analog circuitry, for simplicity and high speed, then one ADC channel will be saved. This may be a more economical approach. An example of an implementation of feedback controlled unmodulated LED power stabilization is shown in Fig. 8.12. The LED output is sampled at PD2 and its photocurrent is compared with a set-point current via R1. If the PD2 intensity increases, the output of IC1 will increase positively. This will reduce the LED current via the current source transistor. The accuracy is now dominated by the temperature coefficient of PD2, which should be a low tempco device such

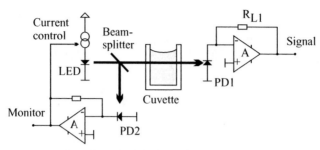

Figure 8.11 One channel of ADC can be saved if the reference is not measured, but used to regulated the source intensity via analog feedback control.

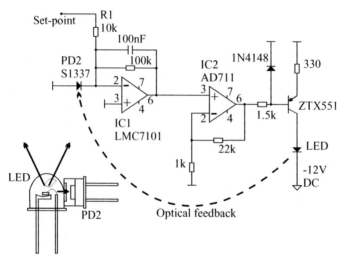

Figure 8.12 LED output intensities can be stabilized by closely coupling source and reference photodiode, e.g., by gluing, and using a simple analog feedback controller.

as the Hamamatsu S1337 series, with <100 ppm/°C. It is sometimes adequate to glue the photodiode directly to the side of the LED where it collects a little of the light, as shown in the insert of Fig. 8.12. Most of the light passes by to the experiment.

In these two-channel measurements it is not strictly necessary to have photodetectors with very low temperature coefficients of output current. It is theoretically only necessary to have two detectors which have *similar* temperature coefficients, and a similar temperature. It is nevertheless highly advisable to use photodetectors with a low tempco. Thermally coupling the two photodiodes will aid temperature tracking and improve precision. It still won't be perfect, but an obsessive approach to matching helps. The use of similar diodes has already been mentioned, meaning the same manufacturer, type, model and package type, batch number if possible, operating temperature, illumination power and density, light wavelength, polarization state, and optical modes. Any detraction from equivalent operating conditions can degrade the tracking of the two receivers' outputs. In the same way the electrical channels should be carefully matched. Large differences in bandwidth, for example between signal and reference channels, will degrade high-speed tracking. It has also been found preferable to synchronously detect both signal and reference intensities. Cost reduction by using an AC-coupled reference path was found to lead to poor performance. The noise contributions of both receiver paths should be understood, as both will contribute to the overall noise level.

There are also many seemingly trivial things that can go wrong. These include mechanical instability and movement of the beam-splitter and photodiodes, vignetting and variable loss of light from each diode, interference fringes on the two detectors, differential contamination of the sample and reference optical paths, and different interfering illuminations. Figure 8.12 shows the simplest type of division-of-wavefront beamsplitter. This is not ideal with semiconductor laser sources due to their significant variation in emitted angular power distribution, which changes with current drive and temperature. Smaller variations take place with LEDs too. It is also important to avoid ambient light falling on PD2. Any power variations here will be mirrored in power variations in the controlled source.

8.5.2   Beam-splitters

The beam-splitter is also a frequent source of problems (Fig. 8.13). The commonest type is the cemented cube, formed using a metal, dielectric or hybrid metal/dielectric thin film on one prism, which is encapsulated against another prism. These are robust and easy to mount, but can have significant polarization sensitivity. Different coating types perform quite differently, so it is worthwhile to study the detailed specifications. Good descriptions of the trade-offs are given in the Newport and Melles-Griot optical catalogs. The accurately parallel surfaces increase the risk of disturbing interference fringes. They can be operated off-normal incidence, but this increases optical aberrations with nonparallel beams.

A simple microscope slide is sometimes pressed into service as a beam-splitter, especially with small diameter free-space beams from lasers. The glass with refractive index n provides a normal-incidence power reflectivity in air of $((n-1)/(n+1))^2$ at each surface, about 4 percent for common borosilicate glass slides ($n = 1.51$). However, the 1mm thickness of the glass plate might be comparable to the input beam diameter, meaning that some overlap of the two reflected beams will occur. With sufficiently coherent sources, this will lead to interference and large relative power fluctuations at the reference detector from the reflected light (Fig. 8.14). As the two reflected beams are of similar intensities, the visibility of the interferences seen there can approach 100 percent. In the transmitted beam interference occurs with a beam that has suffered two

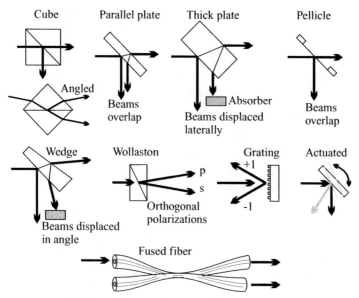

Figure 8.13 Selection of beam-splitters used for intensity referencing.

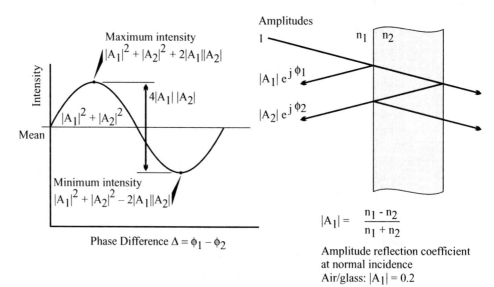

Figure 8.14 Coherent reflection addition in beam-splitters can lead to large intensity split-ratio fluctuations, even with small intensity reflection coefficients.

reflections, and so the visibility is greatly reduced. Calculation of the interference intensity is straightforward. First we write for the two interfering amplitudes:

$$A_1 = |A_1|e^{j\phi 1} \tag{8.3}$$

$$A_2 = |A_2|e^{j\phi 2} \tag{8.4}$$

Where $|A_1|$, $|A_2|$ are the real amplitudes, and ϕ_1, ϕ_2 are the phases of the two waves. In order to calculate the individual intensities, we multiply the amplitude by its complex conjugate:

Intensity of A_1 is:

$$I_1 = A_1 A_1{}^* \tag{8.5}$$

$$= |A_1|^2 e^{j\phi 1 - j\phi 1} \tag{8.6}$$

$$= |A_1|^2 \tag{8.7}$$

Intensity of A_2 is:

$$I_2 = A_2 A_2{}^* \tag{8.8}$$

$$= |A_2|^2 e^{j\phi 2 - j\phi 2} \tag{8.9}$$

$$= |A_2|^2 \tag{8.10}$$

To obtain the combined intensity of the interfering waves, first add the amplitudes to get the total field, then obtain the intensity I as above:

$$I = (|A_1|e^{j\phi 1} + |A_2|e^{j\phi 2})(|A_1|e^{-j\phi 1} + |A_2|e^{-j\phi 2}) \tag{8.11}$$

$$= |A_1|^2 + |A_2|^2 + |A_1||A_2|(e^{j\phi 2 - j\phi 1}) + |A_1||A_2|(e^{j\phi 1 - j\phi 2}) \tag{8.12}$$

$$= |A_1|^2 + |A_2|^2 + |A_1||A_2|(e^{j(\phi 2 - \phi 1)}) + |A_1||A_2|(e^{j(\phi 1 - \phi 2)}) \tag{8.13}$$

$$= |A_1|^2 + |A_2|^2 + |A_1||A_2|(e^{j\Delta} + e^{-j\Delta}) \tag{8.14}$$

where $\Delta = \phi_2 - \phi_1$ is the phase difference between the two waves. Noting that:

$$(e^{j\Delta} + e^{-j\Delta}) = 2\cos\Delta \tag{8.15}$$

we obtain the resultant intensity:

$$I = |A_1|^2 + |A_2|^2 + 2|A_1||A_2|\cos\Delta \tag{8.16}$$

As Δ varies from 0 to 2π, caused for example by temperature variations in the beam-splitter material or small changes in incidence angle, the intensity varies sinusoidally about a mean intensity of $(|A_1|^2 + |A_2|^2)$, with a peak to peak variation of $4|A_1||A_2|$. For example, in the situation of two almost equal-intensity reflections from a glass slide beam-splitter, the intensity will be 100 percent modulated. Even if a 1mW signal interferes with only a 4 percent reflection (0.04 mW) from a single glass reflection, the intensity will vary from 0.64 to

1.44 mW, or about a 55 percent intensity modulation. It can be seen that because it is the amplitude that counts, even surprisingly small interfering intensities can causes large power variations. Effective suppression is therefore very difficult.

An anti-reflection (AR) coating on one surface can be used to reduce the intensity of the second reflected beam. Standard "quarter-wave" AR coatings of magnesium fluoride on glass provide about 1 percent power reflectivity, which sounds pretty good, until it is realized that this means a 10 percent amplitude reflection coefficient. A power reflectivity of less than 0.1 percent can be achieved for higher index substrates, or over a narrow wavelength range by using more complex dielectric stack designs, but this would normally be considered to be very good performance. Even this is 3 percent amplitude reflection coefficient, giving possible interferences with a ±6 percent intensity variation. An alternative AR coating is a textured glass surface. These can also provide power reflectivities less than 1 percent over a broad wavelength range, and very high resistance to optical damage from high power lasers. See Yoldas (1984) for one example. AR coatings are also wavelength- and angle-dependent, more scattering than virgin surfaces, more susceptible to mechanical damage, and difficult to clean. Hence AR coatings should be considered only part of a solution to interference. Interference in detector windows can easily be the dominant source of instability with laser measurements. Avoidance of overlapping collimated beams should be the first consideration. If they must overlap, perhaps the intersection angle can be big enough that many fringes appear across the detector. Last, if the application can work without a high-coherence source, for example by substituting a multimode semiconductor laser, then achieving stability will be much easier. This is one approach taken in optical disk readout optics.

A thick glass plate is an alternative beam-splitter. With the larger lateral separation of the two beams, overlap is reduced and physical interception and absorption of the unwanted beam is easier. The only problems likely here are increased optical aberrations and scatter caused by the thick glass, beam-shifts, and perhaps extra cost.

As thick plates are one approach, very thin beam-splitters ("pellicles") are another offered by several manufacturers. These are membranes of polycarbonate or other tough plastic a few microns thick and held in tension on a metal frame. The small thickness means that the two reflected beams are fully overlapped, and most sources will suffer interference, but the small optical path difference (OPD a few λ) causes little change in OPD with wavelength or temperature. Hence there should be no intensity variation. Perhaps the biggest problem with pellicles is their high compliance; they tend to act as microphones unless carefully protected against air currents, sound and vibration. The smallest pellicle which will accept the beam should be chosen, although the majority of products are large ($\approx$25 to 50 mm diameter). Five millimeter pellicles would be much stiffer, and more useful for much of modern optoelectronics. Perhaps someone supplies them. Perhaps there is a market too for single crystal pellicles, formed from pure or doped silicon and other semiconductors, as less compliant pellicles for near infrared beam-splitters.

The optical manufacturers also offer glass plates polished into a wedge with a small angle, (for example 5 minutes of arc). In this case the two reflected beams are separated in angle, as well as in offset. The angle gives interferences with much finer fringes which can average out over a large detector's area, and makes it easier to block and absorb the unwanted beam at a convenient distance from the beam-splitter. The main disadvantages are the shift in the transmitted beam optical axis and the aberrations.

For fiber-connected sensor systems, fiber-couplers are very attractive as beam-splitters for optical referencing. The primary, and large, problem with multimode fiber couplers is that they are notoriously sensitive to the distribution of modes entering the coupler. In some types the low-order modes preferentially continue on along the input fiber strand, while higher order modes couple across to the second fiber. Unless effective mode mixing and power-equalization is in place large fluctuations in split-ratio are to be expected, except in the most benign of environments. This is the case for both incoherent and laser sources. One big advantage of fiber splitters is their completely closed path and hence freedom from contamination.

Where a coherent (laser) source is used, speckle effects will also lead to split-ratio fluctuations (modal noise). The speckle pattern present at the point of coupling will be sensitive to even small movements of the input fiber. The solutions here are to either substitute a source of lower coherence length (e.g., an LED), or to dynamically vary the speckle pattern though the use of a fiber shaker or moving diffuser at the point of coupling into the fiber. These potential errors are in addition to the severe intensity fluctuations seen if a multimode fiber output is vignetted before detection. As speckles move on and off the detector with fiber movements, large intensity fluctuations can be caused. In my experience, the loss of signal intensity in changing from a laser to an LED in fiber-based systems can often be more than compensated for in the reduced speckle interference effects. This even includes the use of low-power visible LEDs coupled to single-mode fibers (see the TRY IT! in Chap. 3). Last, both single-mode and multimode types are available, in silica, glass, and plastic. All fiber components exhibit polarization sensitivity, which are more pronounced due to little averaging in single-mode fibers.

Where small fractional power tapping is adequate, fiber-bend taps can be very convenient. Here the fiber is subjected to a small bend radius by bending or wrapping it around a small mandrel. It is generally not necessary or desirable to strip the fiber, but to clamp onto the primary or secondary coating. The radius of curvature may be changed to adjust the output coupling efficiency. However, strong out-coupling can sometimes take the fiber close to its minimum allowed bend radius, so care is needed. Clamp-on sources and detectors are available commercially, which are very convenient for a range of optical test and measurement applications. One home-made design which is easy to make is shown in Fig. 8.15. This uses a plastic *heat-shrink* coated steel mandrel to bend the

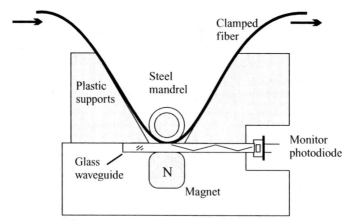

Figure 8.15 Light can be tapped from a fiber using a small-radius bend. A magnet and steel mandrel make a convenient clamp, and even allow the light to be coupled to an embedded photodiode.

fiber with a known radius of curvature and to press it against a glass baseplate. If a primary-coated fiber is used, light couples out of the coating and into the glass slide at the point of contact. The glass plate acts as a simple light-guide coupling to a small photodiode such as a BPW34. The steel mandrel can be held in place using a small magnet fixed underneath the glass.

Schemes for optical referencing in these ways provide some of the highest resolution measurement techniques available. The papers by Allen (1995) and Imasaka (1983) give good examples of what is possible in chemical photometry.

8.5.3 Intensity noise reduction

Dual-detector referencing has another very important use—source noise reduction. In systems that have high received optical powers, which should easily be shot-noise limited in detection, the measured noise level is often found to be far greater than expected from the shot calculation. This is because the source itself exhibits excess intensity noise. Sometimes this is due to poor regulation of the power supplies feeding the source. This can usually be seen with a low-frequency spectrum analyzer. Even with good supplies, laser systems in particular are prone to excess noise, usually exacerbated by spurious optical feedback into the laser cavity from lenses, windows, fibers etc. Even focussing the laser beam with a microscope objective onto a cleaved fiber end-face for coupling forms an efficient retroreflector. The noise level is often 30 to 60 dB worse than shot noise. Just as we can suppress slow intensity drifts caused by source temperature fluctuations, so we can use the configurations such as seen in Fig. 8.10 to reduce the effects of fast excess intensity noise. For example, we can divide signal by reference in an analog divider IC as previously mentioned (Fig. 8.16a). Alternatively, if the two signals are adjusted to be nominally equal, they can be subtracted to achieve the same compensation (Fig. 8.16b). This is capable of lower noise and higher bandwidth.

(a)

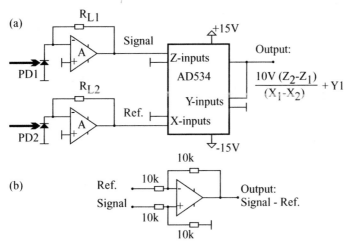

(b)

Figure 8.16 Analog division of signal and reference is straightforward, but limited to about 1 percent precision, even at low frequencies. If signal and references are equal in magnitude, subtraction can offer higher precision and greater bandwidth.

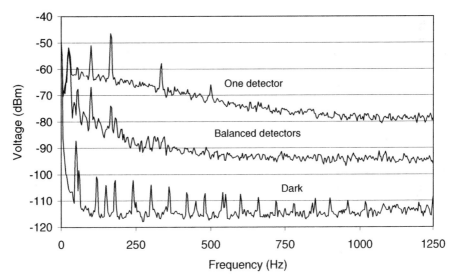

Figure 8.17 Noise spectra of a $1.55\,\mu m$ diode laser source illuminating just one detector, or two detectors arranged for equal photocurrents. The shot noise level at this photocurrent is $-98\,dBm$.

Figure 8.17 shows intensity noise from a $3\,mW$ $1.55\,\mu m$ distributed feedback laser, detected on an InGaAs photodiode and $1\,M\Omega$ transimpedance. With $5\,V$ output ($5\,\mu A$ photocurrent) the shot noise limit should be $1.27\,\mu V/\sqrt{Hz}$, or $-98\,dBm$ in the $4.6\,Hz$ bandwidth used. The traces show the measured low-frequency noise level with the photodiode masked (lowest curve), and with the laser light received on one detector (top curve). It can be seen that the noise level over most of this low frequency region is 20 dB higher than the shot limit. The central curve shows the result of subtracting an almost identical reference current obtained from the same beam with a beam-splitter from the signal photocurrent. The two received powers have been manually adjusted to almost null out the DC photocurrent. If this were done with two shot-noise limited signals, the uncorrelated noise powers would add, increasing the measured noise power by 3 dB. In fact, the noise drops to within 4 dB of the shot noise level. The even lower dark noise shows that amplifier noise is insignificant, even with visible power supply harmonics.

One advantage of performing this current subtraction directly with the photocurrents using series-connected photodiodes as in Fig. 8.18a is that there are no bandwidth limiting elements in place, so the noise reduction should be effective over the full bandwidth of the photodiode. Otherwise it depends on the *dynamic* balancing of the two receiver channels, which is difficult to achieve to high precision. Houser (1994) has used this technique to make 0.1 percent precision optical transmission measurements in megahertz bandwidths. The problem with subtraction is that manual adjustment of the two signals must be very precise, which is difficult to arrange and maintain. Philip Hobbs (1997) has published a highly elegant and capable optoelectronic module for performing the photocurrent subtraction. This is equivalent to the simple current subtraction of Fig. 8.18a, except that the fraction of reference current used can be adjusted electronically in the differential transistor pair. With the addition of an electronic feedback loop to automatically adjust the current subtraction, the circuit fragment of Fig. 8.18b becomes the basis of a very high performance intensity compensation system. With care to match the two optical signals by

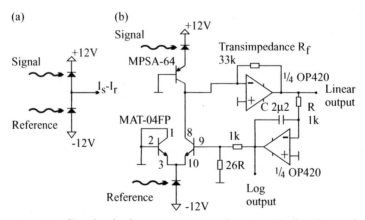

Figure 8.18 Signal and reference currents can be automatically subtracted using this elegant circuit fragment. A feedback loop adjusts the proportion of the reference photocurrent that is subtracted from the signal to give zero. This allows intensity noise suppression right down to the shot noise limit. Circuit reproduced by permission of P.C.D. Hobbs.

making sure they come from the same part of the source beam, in the same polarization state, free from asymmetrical intensity fluctuations caused by interference fringes in windows and other components and so on, the effective measurement noise level can be brought down to within a decibel of the shot noise limit over a wide bandwidth. This constitutes an enormous improvement. In one stroke a large number of laser sensor systems are turned from offering marginal performance to delivering real precision.

8.5.4 Split detectors

Another useful application of optical referencing avoids the discrete beam-splitter altogether, making use of different parts of the source beam as in Fig. 8.12. High resolution position sensors can be constructed by projecting a high intensity beam onto a split photodetector. Figure 8.19 shows two photodiodes made in a single substrate. The structure has common diode anodes. An opaque aperture is used to cast a well-defined spot of light from an infrared LED onto the junction between the two sensitive surfaces. If the spot is symmetrically placed, the signal formed from the difference between the two photocurrents $V_L - V_R$ is zero. With a displacement right or left we have a positive- or negative-going signal.

With an optimized choice of the spot-diameter the transfer function can be adequately linear for simple position sensing applications. A square aperture of uniform illumination would improve linearity, but even a bare LED placed close to the photodetector can be used (lens removed by polishing). The sensor gain or slope of the transfer function then depends on the axial separation Z. Gain stability is improved through use of the aperture to restrict the angular diver-

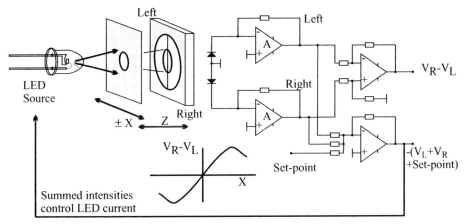

Figure 8.19 Very simple but high resolution 1- and 2-dimensional position sensors can be fabricated using split (dual or quadrant) photodiodes illuminated by light from an LED, preferably cleanly apertured. The photocurrent difference gives the position, while the sum can be used to control the source intensity. Remanent scale factor errors occur if the axial separation Z changes.

gence of the beam. The resolution of this sensor clearly depends on the transfer slope, and this can be made steeper through use of a smaller illuminating spot. However, if the spot diameter becomes of the order of the dead-region width between the two photodetectors, linearity will suffer. Two- and quadrant-cell detectors are available from Centronics, Hamamatsu and others with dead-zones typically $10\,\mu m$ to $100\,\mu m$ wide. As the total received power can be high, very high shot-noise limited position resolution is possible. With $100\,\mu A$ photocurrents and a 1-mm wide linear region the shot noise limited resolution is about 5.7×10^{-8} of 1 mm in 1 Hz bandwidth, or 57 pm. Even in a 10 kHz bandwidth required for a fast mechanical position servo mechanism this is 5.7 nm rms. We will at least lose a factor $\sqrt{2}$ because of the uncorrelated noise contributions of the two photodiodes, and amplifier noise and bandwidth limitations must be evaluated as in all our designs. Nevertheless, this is a pretty impressive resolution for such a simple device. It is the basis of many readout schemes for the atomic force microscope. There, a laser beam is reflected off the probe cantilever to impinge on a split detector. Resolution can be increased by increasing the received laser power, although local heating can be a problem.

Position accuracy of the basic scheme is not as impressive as resolution. We have mentioned geometrical gain-changes caused by beam-diameter variations, and we have the old problem of LED and laser output power variation. This can be corrected by using the optical referencing options in-built to the configuration. As long as the beam-shape is constant, the straightforward division of $(V_R - V_L)$ by $(V_R + V_L)$ will correct for LED intensity. Accuracy will be no better than the divider used; about 1 percent for an analog divider. Digital division accuracy will be limited by the ADC resolution used, and if both signal and reference voltages are close to full scale, most of the resolution of a 12- or 16-bit ADC could be used. Signal reductions will of course quickly reduce the number of effective bits of resolution. Higher resolution ADCs are available, especially the "$\Sigma\Delta$" devices which extend out to 24-bits. However, increasing resolution reduces sample rate, and achieving such resolutions for the whole measurement is very difficult.

An alternative approach is to use light-regulation techniques such as in Fig. 8.12, with the summed intensity stabilized. Again, we save one ADC, and in most cases achieve higher resolution and accuracy by avoiding division. The same approach can be used for two-dimensional positioning using a quadrant-cell.

8.6   Multibeam Referencing

Simple power referencing relies on the precision tracking of detection sensitivity of the two (signal and reference) photodiodes. Over the life of a product, this may be difficult to guarantee to the required degree of precision. In principle, even this error can be eliminated with multibeam referencing.

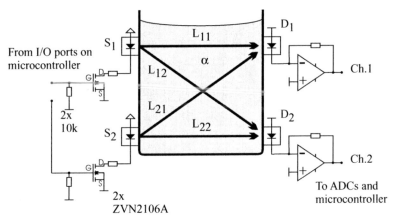

Figure 8.20 Four-beam configuration. By measuring transmission intensities along the four paths L_{ij}, the absorption coefficient of a uniform material can be determined even with slow changes in source output and detector responsivity.

8.6.1 Four-beam referencing

Figure 8.20 shows a "four-beam referencing" system, widely used in fluid absorption measurements. Two sources (e.g., LEDs) and two detectors are used. The LEDs and photodiodes have sufficiently large emission and acceptance angles to "see" each other. By alternately switching the two LEDs on and off, the transmission of all four paths L_{ij} shown can be determined. In the case of a medium with uniform absorption coefficient (α), the transmitted signal powers due to source i and detector j are of the form:

$$P_{ij} = S_i D_j e^{-\alpha L_{ij}} \tag{8.17}$$

where S_i is the source output power, D_j is the detector sensitivity, and L_{ij} is the path-length of path ij. By forming the function Q, the variables S and D can be eliminated:

$$Q = \left[\frac{P_{11}P_{22}}{P_{12}P_{21}} \right] \tag{8.18}$$

$$Q = \left[\frac{S_1 D_1 S_2 D_2}{S_1 D_2 S_2 D_1} \right] e^{-\alpha(L_{11}L_{22} - L_{12}L_{21})} \tag{8.19}$$

$$Q = e^{-\alpha(L_{11}L_{22} - L_{12}L_{21})} \tag{8.20}$$

This last expression looks like a conventional transmission experiment with an absorption coefficient α and effective length of $L^* = (L_{11}L_{22} - L_{12}L_{21})$. The source powers and detector gains have dropped out of the expression completely! This configuration is therefore widely used to determine the absorption and scat-

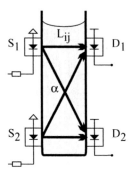

Figure 8.21 Resolution of the four-beam configuration is maximized by increasing the difference between "short" and "long" path lengths.

tering attenuation of liquids in water treatment processes. Even without stabilization of the source powers, and without correction for sensitivity changes in the receivers, the technique can be used to determine α with good stability. To maximize sensitivity we must maximize the difference between diagonal and straight-through paths (Fig. 8.21), and hence L^*. However, the wide source beam dispersion and detector acceptance angle are difficult to arrange, and make the system more prone to fouling errors. Note that the approach is not perfectly independent of S_i, D_j, as any contamination seen on the surfaces of the optoelectronic components is unlikely to be identical for the straight through and diagonal paths. These problems are discussed further in Chap. 9. Nevertheless, the technique is very elegant and useful in practice too. In small sensors, the stability of this configuration is dominated by reflections off the container walls. We have also built large systems with $\approx$1m paths for UV gas-phase sensing using pulsed xenon lamps. Here we have the additional problem of capturing the peak pulse energy after the two photoreceivers using a sample-and-hold amplifier, and in matching the dynamic performance of the two receiver channels. Where low-pass filtering is needed, for example using Bessel filters instead of Butterworth filters for their superior pulse response, both channels must be carefully equalized. If this is done, stabilities approaching 16-bit precision have been achieved.

8.6.2 N-beam systems

The use of multiple beams and sufficient independent measurements to calculate all the free variables S_i, D_j begs the question: Could we do even better with more sources and detectors, and more than four beam-paths? This depends on the path-length differences we can obtain. If Fig. 8.20 is arranged with sources and detectors in a "square" pattern, $L^* \approx \sqrt{\text{cell dimension}}$. Alternatively we could use four sources and four detectors with 16 possible paths, which could also be arranged to cover a square pattern. In this case (Fig. 8.22) $L^* \approx 1.53 \times (\text{cell dimension})$. The gain in effective path-length is small, but the extra degree of averaging the 16 beams may be useful.

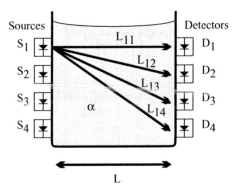

Figure 8.22 More sources, detectors, and path permutations can improve absorption resolution.

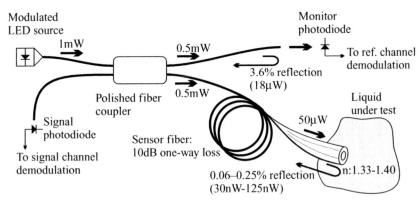

Figure 8.23 Measurement of changes in power reflected back from a distant fiber end is attractive for a wide variety of sensing applications. However, performance is severely compromised by changes in fiber loss, or by the more intense reflection from a monitor-photodiode branch. The simple configuration is unusable except in the most benign environments.

8.7 Design Example: Multimode Fiber Refractometer

Figure 8.23 shows a basic reflection-mode refractometer constructed using multimode fibers. This is an initially attractive system for detecting and measuring small refractive index changes in a liquid, brought about by chemical concentration changes. Similar systems have been used in applications such as a remote saccharimeter for sugar concentrations and sensors to detect the state of charge of secondary batteries. The idea is to use the variation of power reflection coefficient at the bare-fiber/liquid interface to give an intensity change related to the liquid refractive index. The reflection coefficient is given by the Fresnel relations.

The source chosen is a high intensity near-infrared LED which couples 1 mW into the 200 μm core diameter, 0.3 numerical aperture (NA) silica-on-silica fiber, whose loss is 1 dB/m. In order to compensate for coupled power variations of the source, a polished fiber coupler is used to tap off one-half of the source power to a high-quality connectorized monitor photodiode with negligibly small temperature coefficient. The signal fiber is likewise connectorized and coupled to a receptacle-mounted photodiode and transimpedance receiver. Responsivity of both diodes at 0.88 μm wavelength is 0.6 mA/mW.

The distal end of the probe fiber is 10m away, immersed in the aqueous process liquid, whose refractive index n_2 varies from 1.33 to 1.40, roughly corresponding to 0 to 40 percent concentration on the Brix scale. Although there is a range of incident angles of light within the fiber core passing into the liquid (approximately $\pm\sin^{-1}(\text{NA}) = \pm17°$), we will just use the normal-incidence Fresnel relations, where the power reflection coefficient R is given by:

$$R \cong \left[\frac{n_1 - n_2}{n_1 + n_2}\right]^2 \tag{8.21}$$

Hence with an average fiber refractive index of $n_1 = 1.47$, the power reflectivity at the fiber end varies from 0.06 percent to 0.25 percent in the different process liquids, and the signal power can then be calculated from the double pass through coupler and sensor fiber. We obtain for the detected signal power:

$$(n_2 = 1.40)\text{: } 1\text{ mW} \times 0.5 \times 0.5 \times 0.1 \times 0.1 \times 0.0006 = 1.5\text{ nW}$$
$$(n_2 = 1.33)\text{: } 1\text{ mW} \times 0.5 \times 0.5 \times 0.1 \times 0.1 \times 0.0025 = 6.25\text{ nW}$$

The monitor channel detector should generate a photocurrent $I_r \approx 400\,\mu$A, so with a transimpedance amplifier load of 12k we have a voltage or 4.8 V, ideal for digitization in a 0 to 5 V ADC. This channel should provide few difficulties. The first problem with the sensor configuration is the relatively low received power at the main signal photodiode. With 6.25 nW power and 3.75 nA maximum photocurrent, we need a transimpedance of 1 GΩ for 3.75 V output. As we don't have any, we will use a 100 MΩ transimpedance and a further stage of 10× gain. With a minimum of 90 mV at the transimpedance output, we are likely to be just shot-noise limited.

The modulation frequency is chosen as 3892 Hz, well above the main interference sources, but not so high that achieving adequate bandwidth is difficult. The level of detected ambient light in the process chamber will be negligible, and as the light is fiber-guided, a small, low-capacitance photodiode can be used to capture it and achieve 10 kHz bandwidth. Much more worrying is the reflection from the monitor photodiode assembly. If this is a simple polished fiber connector butt-coupled to the photodiode, the fiber end will reflect about 3.6 percent of the incident light, contributing another received signal power of the order of:

Reflected monitor power: $1\text{ mW} \times 0.5 \times 0.5 \times 0.036 = 9.0\,\mu$W

This is three orders of magnitude larger than the peak signal from the process liquid, and modulated at the same frequency! In order to resolve even ten refractive index values with this system the total stability of all reflection coefficients and fiber attenuation values would have to be about 20,000:1. This is not impossible, but represents an extremely high performance system.

Clearly the sensor fiber loss is a big problem, so we might attempt to find an alternative fiber with lower loss. However, even if the sensor fiber loss were reduced to zero, the monitor diode reflection $(9.0\,\mu W)$ would still dominate the maximum primary received signal (now 625 nW). Hence the next job is to reduce the monitor diode reflection, and we have several options. Antireflection coatings in the form of quarter-wave thickness films of geometric-mean refractive index deposited on the fiber end can in principle reduce the reflection to zero. In practice, restricted choice of coating materials, errors in deposition, the refractive index gradient across the fiber core diameter and the spread of propagation angles in the coating conspire to reduce the coating performance. We might expect 0.25 percent power reflection coefficient, or about 625 nW at the signal photodiode.

Better anti-reflection performance can be obtained using carefully matched transparent gels or angle-polished fiber ends (Fig. 8.24), after Ulrich and Rashleigh (1980). By choosing the angle correctly, reflected rays are not recaptured in the acceptance numerical aperture of the fiber. With single-mode fibers of N.A. ≈ 0.1 an angle of 12° can give 10^{-6} power reflection coefficient, reducing the spurious reflected power to 0.25 nW and give about 20 resolved refractive index values. At last the design looks like it might just measure something, but it is still not exactly robust. With higher N.A. fibers the angle must be increased, making polishing and coupling much more difficult.

Perhaps the coupler is the problem. What about asymmetric couplers with only a 1 percent tap-off fraction? The disadvantage here is that although a 99 percent cross-over coupling ratio from source to sensor is a big help, returning light also crosses efficiently, propagating back to the LED source. There are ways to use asymmetric couplers to our advantage, but it is usually necessary

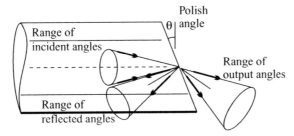

Figure 8.24 Reflections from fiber ends and diode laser windows are routinely reduced by angle-polishing. For good performance the cone of reflected light must not overlap the acceptance cone of the fiber. This becomes more difficult as the numerical aperture increases.

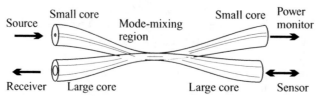

Figure 8.25 Asymmetric power splitters, for example formed using two types of optical fiber, can reduce the effects of reflections in the power monitor branch.

to use different fiber types. A small core and a large core fiber fused together to effectively mix modes can help (Fig. 8.25). Here the input light passes primarily to the sensor fiber, but returning light passes primarily to the signal detector. Only a small fraction is tapped off to the monitor photodiode. This sounds perfect, except that the small source fiber also reduces the power coupled from a multimode source such as the LED. If we allow ourselves loss, then a simpler approach is to carefully splice a long, lossy length of fiber between coupler and monitor photodiode. With 20 dB one-way loss the monitor signal will still be adequate for low-noise detection, but the 40 dB two-way loss will significantly reduce spurious reflections, especially if combined with coatings, gels or angle-polishing.

In Chap. 11 we will discuss additional electronic measures which can make further improvements. In the interim the simple design points out the very great difficulties of intensity-based measurements of this sort. While it is not impossible to build the fiber reflectometer for refractive index measurements great care is needed to make it effective even in a benign environment. For the "real world" there are probably much better solutions.

Contamination and Industrial Systems

9.1 Introduction

The multibeam referencing and intensity compensation configurations introduced in the last chapter are of enormous help with another problem that affects the long-term stability of an optical measurement. In this Chapter we will think about some of the problems involved in transferring a high performance optical instrument, well designed in consideration of photon budgets and temperature drift performance, to an industrial environment where it is up against a different set of stresses, the greatest of which is so called *fouling*. This is the inevitable contamination that builds up on the optical surfaces and windows of an instrument in service. Several of the techniques presented here are related to those used for stability improvement in Chap. 8, although the emphasis is more on environmental effects rather than temperature variations. We will look at some real measurement problems, and the designs of a few instruments which have been used successfully in difficult industrial sensing applications.

9.2 Transfer to the Industrial Arena

Large numbers of optical sensors used for industrial diagnostics and control systems operate on-line (i.e., a sample of fluid is brought to the instrument in dedicated pipe-work) or even in-line (where the sensor is immersed in a flowing process fluid stream or installed in a section of process piping). In both cases the optoelectronic elements, sources and detectors, are separated from the fluid by robust transparent windows, for instance of plastic, glass, fused-quartz (amorphous SiO_2) or sapphire (single-crystal Al_2O_3). Although it is often assumed that the window is just a passive, protecting interface between sample and sensor, over time it can become a significant variable element in its own right. In particular, windows can become coated with chemical deposits, often aided by biological films. As the fouling layer builds up, optical transmission decreases and fluorescence can increase, leading to measurement errors.

In some processes designed to precipitate chemical species, such as the manganese-removal stages of drinking water treatment, highly opaque, physically hard and adherent films can build up on instrument windows in a few hours (in this case black, highly absorbing films of manganese dioxide). Even in clean drinking water supplies, nutrients and oxygen are always present, allowing bacteria to grow and form "biofilms." Again, this growth can be a fast process. Ultrasmooth window surfaces of polished sapphire or highly inert materials such as transparent PTFE and other fluoropolymers may delay the onset of film growth, but once a few bacteria have taken hold, further growth is often rapid. Films can range from predominantly transparent, sticky polymers to hard, opaque layers with combined inorganic constituents from the fluids themselves and from corrosion products of the pipe-work.

Reference-beam configurations, and in particular the four-beam approach repeated in Fig. 9.1, are widely used to counter window fouling in these industrial sensors. We calculate the ratio of "short" to "long" path intensities or scattered to transmitted intensities:

$$Q = \left[\frac{P_{11}P_{22}}{P_{12}P_{21}}\right] = e^{-\alpha(L_{11}L_{22}-L_{12}L_{21})}$$

which appears to be independent of source and detector window fouling. The basis of the intensity compensation scheme is that the intensities measured along paths L_{11} and L_{12} are perfectly proportional, no matter what is the level of fouling, and the compensation scheme should be perfect. In practice, the

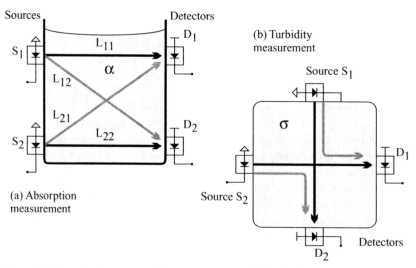

Figure 9.1 Four-beam compensation systems are used for (*a*) absorption and (*b*) scattering measurements.

system does help, and transmission reduction to 10 percent of that of the clean state can usually be encompassed. Nevertheless, the reduction in light will certainly reduce the S/N and increase errors due to the numerical division processes involved in compensation. The technique is used both for liquid absorption measurements using four transmission determinations, as well as for turbidity measurements with determination of two transmitted and two scattered intensities. Other combinations and geometrical arrangements are possible. The electronic division limitations are not fundamental, and could in principle be lifted through higher resolution ADCs, brighter, self-adjusting sources, etc. We should really be asking what are the remaining fundamental limits of the technique which preclude perfect compensation?

The first consideration is geometrical (Fig. 9.2). Plane window devices with different path directions will always show different attenuations for the P_{ii} and P_{ij} signals, as the path-lengths in the fouling layer are different for the two beams. As the fouling layer builds up, the tracking of the signals in the two directions will diverge, leading to errors. We could attempt to improve things by using hemispherical windows with a radius of curvature much greater than the film thickness (Fig. 9.3), centered on the source/detector, in order to give similar layer thicknesses and hence attenuation in the two paths, but the advantages are likely to be eroded by the increased lateral separation between the two effective windows and by the increased risk of biogrowth compared with a plane window.

Second we must consider the structure of the contaminating films. Even if the geometrical effect is weak, we still require that the film is uniform. Any contaminant on the window close to source S_1 must affect both beams identically, which depends on the details of the fouling layer. Errors will be seen if the deposited film thickness is nonuniform, or there exist absorption nonuniformities. If the fouling film shows granular variations on a scale much larger than source and detector, the compensation should still be effective. Likewise,

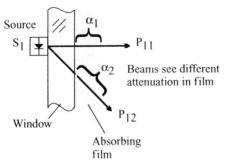

Figure 9.2 Fouling layers are not completely compensated by the four-beam technique, as the different beams see different layer thicknesses and attenuations.

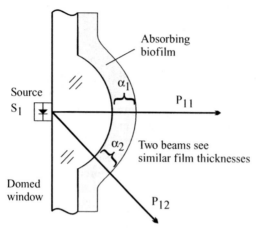

Figure 9.3 Domed windows can equalize the absorbing film atternuations, but at the risk of greater speed of contamination growth.

if the deposited films vary on a scale much finer than source and detector, then the compensation should also be effective. The real problems arise where the fouling is structured at just the *wrong scale*. We could say that the density of fouling variations with spatial frequencies of the order of the reciprocal of source/detector dimensions should be low to obtain good compensation. Stated less pretentiously, dirty blobs about the same size as source or detector (Fig. 9.4a) are bad news! They cause the attenuation to be different for the straight-through (P_{ii}) and cross (P_{ij}) beams. One help is to make the sources and detectors much larger than the typical blob size (Fig. 9.4b). In pumped-sample on-line instruments with small cells, the options for this are rather limited. However, where a four-beam attenuation measurement system is to be immersed in the large scale flowing water stream typical of a water treatment works, size is hardly a consideration. There is plenty of space and sample available. It therefore seems surprising that no one, to my knowledge, manufactures instruments with sources the size of car headlamps, and solar-cell detectors almost as large. Lamps made up from dozens of light emitting diodes are becoming common for illumination and display applications, which would be ideal for such a large-format turbidity or optical absorption instrument. They could still be modulated at high speed for synchronous detection, and are likely to provide much more robust measurements, both in the face of severe contamination build-up and with nonuniform samples such as raw sewage and effluents.

The basic four-beam configuration of Fig. 9.1 is probably about right for many applications, aided by optoelectronic components chosen with regard to the structure of the fouling layers expected, and especially if helped by occasional automatic cleaning. Although perfection is unlikely, some form of fouling compensation using multibeam referencing with plane windows should be designed in to all but the most basic instrument.

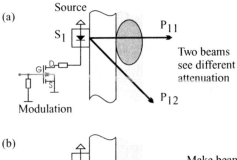

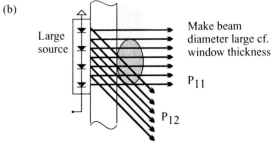

Figure 9.4 Contamination of just the "wrong size" reduces the effectiveness of multibeam compensation (*a*). Sources and detectors which are large with respect to the contamination granularity can help (*b*).

9.3   Wavelength Referencing

9.3.1   Two-wavelength measurements

Instead of using two physically separated beams for signal and reference, we can often use two beams of different wavelength. This approach is widely used in analytic chemistry where a chemical species is to be detected which absorbs at one convenient wavelength and not at another. Using beam-splitters, fiber couplers, interspersed fiber bundles, or even just an array of LEDs, it is possible to have the two signal beams overlap almost perfectly at the windows, such that they see very similar averaged contamination and attenuation (Fig. 9.5). Multiple interspersed sources at the two wavelengths help with this equality. The simplest situation is where the contamination shows no spectral character, that is, it looks grey and colorless. The spectral feature of interest just sits on top of a uniform background (Fig. 9.6, here a weak solution of the indicator M Cresol). As the fouling builds up, the total absorbance spectrum increases and perhaps becomes more noisy, but the height of the absorption feature remains constant. Measurements made on the peak of the known absorption and off to one side can therefore be used to compensate for variable contamination attenuation. Simply by forming the ratio of the signal and reference intensities we obtain the relative attenuation due to the spectral feature. For example, if the intensity at the wavelength of peak absorption is 90 percent

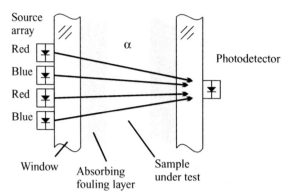

Figure 9.5 Multiple wavelengths, preferably with beams interspersed to equalize attenuation, can reduce the effects of contaminating films.

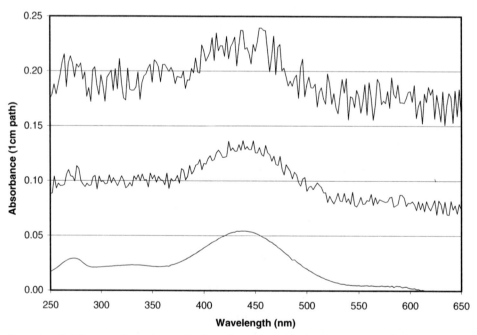

Figure 9.6 A full spectral measurement of an absorption of limited extent can ease determination of its magnitude, even in the presence of large uniform absorbance and an increased noise level.

of that off the absorption feature, this ratio of 9/10 ($A = -\log_{10}(9/10) = 0.046\,\text{AU}$ absorbance units) will be obtained regardless of the multiplicative attenuation due to the contamination. Alternatively, on an absorbance scale we must determine the difference by subtracting the on- and off-peak absorbances. The effect of the fouling is just a vertical offset of the chemical species' spectrum, and the

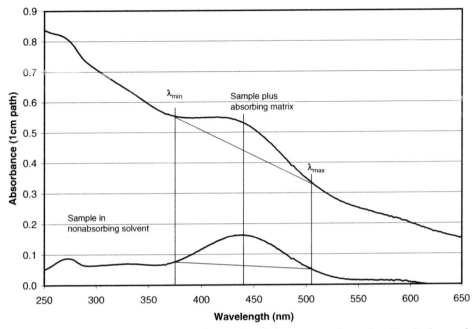

Figure 9.7 Where the known absorption feature is added to a nonuniform absorbing background, three-point measurements can help to recover the true absorbance. Calculate at the peak, or find the area under the feature.

absorbance difference will remain fixed. Hence, even if the attenuation due to the contamination changes with time, the dual-wavelength measurement will be well compensated.

9.3.2 Three-wavelength measurements

In many cases with both chemical and biofilm contamination the film attenuation is not spectrally flat, and so the two-wavelength procedure breaks down. Now three wavelengths will do a better job of estimating the true species' absorption (Fig. 9.7). A common procedure is to draw a straight line between the absorbance values at the lower and upper reference wavelengths (λ_{min}, λ_{max}), equally spaced about the peak absorption. Then we can reconstruct the baseline and determine the absorbance at the peak λ_c from the elevation at the peak, or integrate the area under the absorbance feature between the bounding wavelengths. In this case the absorption spectrum of the contaminant can change in magnitude and in slope without greatly increasing the error. If a good model or measurement of the contamination alone is available, for example by measuring a pure water "blank" in the contaminated cell, then the accuracy of the integration can be further improved. Unfortunately, the spectral absorption of fouling is often neither well characterized nor predictable, so errors will always be present if using a model.

9.3.3 Turbidity correction

A widely practiced instance of this compensation is found in water absorption measurements. When measurements are performed at blue wavelengths, large errors are often caused by attenuation from scattering from particles. The variation of the attenuation due to scatter with wavelength is quite repeatable, as long as the size and shape distribution of scattering particles is constant. By measuring at short and long wavelengths (typically 470 nm and 880 nm), some correction for the turbidity is possible. At 880 nm there will be attenuation due to particle scatter, but negligible absorption. At 470 nm there will be both absorption and scatter. The 880 nm attenuation measurement can therefore be used to estimate the attenuation due to scatter at 470 nm, and subtracted from the total attenuation, leaving just the 470 nm absorption. Compensation is not perfect, but is a help with samples with relatively well-known and stable properties.

As an aside, note that even pure, particle-free liquids show some light scatter, due to temperature and density fluctuations at a molecular level. Far from being negligibly small, it is easy to see this turbidity by shining a laser pointer through a glass of clean liquid in a darkened room. No matter how pure is the sample chosen, the scatter is visible to the eye. It also limits the lowest measurable turbidity of a continuously measuring instrument to about 0.02 NTU (nephelometric turbidity unit). (Good quality drinking water has a typical turbidity at production of 0.1 NTU). Lower turbidities can be measured by restricting the volume of water sensed, whereby the signal is detected via the change in intensity as a particle moves through the detectors field of view. This is an example of sensitivity enhancement due to *measurand modulation* described in more detail in Chap. 10.

9.3.4 Multiple-wavelength measurements

Knowing the target spectrum of the indicator of the lower curve of Fig. 9.6, it is quite easy to see the same feature in the uppermost curve, despite the poor S/N. Common sense suggests that the more interesting features there are in the target species spectrum, and the less in the contamination, the easier it will be to detect the target, and the better will be the compensation for fouling. That is, we should use as many wavelength channels as possible. This is a useful approach in detection of some colorimetric indicators, as they often exhibit a series of well-defined absorption peaks across the visible and UV regions. Then a selection of optical sources placed on the known absorption bands, and another selection arranged to measure between the absorptions, suitably divided, will improve compensation for (slowly) varying spectral characters of the fouling (Fig. 9.8). In practice, with any more than four or five sources, it may be more convenient to collect a full spectrum of 256 or more wavelength "bins" using one of the miniature spectrometer modules from Zeiss, Ocean Optics, Microparts and others. Selection of on-band and off-band wavelength channels, together with all fouling compensation, can then be done in software. Optimally choosing the on- and off-band wavelength regions is tricky by hand. There are formal

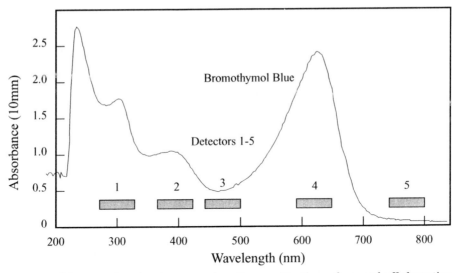

Figure 9.8 More complex absorbance spectra allow combinations of on- and off-absorption detectors to give improved suppression of background absorption.

tools for this generally known as linear multivariate calibration schemes, such as multiple linear regression (MLR), principal component analysis (PCA), and partial least squares regression (PLS). If these linear systems fail to give good calibration because of nonlinearities of absorbance versus concentration, it may be useful to try neural nets, which can provide similar calibration functions but also deal with nonlinearities. Commercial spectral analysis software is available with all these tools included, or you can develop your own using the available mathematical descriptions and general programming software. Martens and Næs (1993) give a very clear and practical description of the linear techniques.

9.4   Window Cleaning

Fouling should never be considered as an afterthought to a manufactured instrument, as though we might get away without dealing with it! Fouling always occurs, and on-line instruments have a tendency to be installed and forgotten. Even laboratory instruments will suffer from similar types of fouling, but the shorter periods actually in service, the proximity of staff, and the greater ease of cleaning give fewer problems. Nevertheless, any instrument should be designed from an early stage with fouling compensation, removal, or avoidance in mind. Let's now look at the second of these; removal means cleaning or replacement.

9.4.1   Disposables

The simplest approach for many measurements is to use disposable windows. Usually this means a whole disposable sample cell, for example a plastic cuvette with molded windows. Polystyrene and acrylic plastics are the most common materials for these, which are now available with very high quality optical surfaces at low cost. Their wavelength of operation is limited at short wavelength to approximately 350 nm, but for many laboratory, treatment works, and industrial manual testing applications using spectroscopy, turbidity, and indicator chemicals, this is sufficient. Although they are often used for repeated measurements, the plastic's easily damaged optical surfaces make this a risky endeavour compared with glass or silica cuvettes. In any case, the cost of these measurements is usually dominated by the cost of getting to the site, obtaining a sample, performing the measurement and thinking about the results, so that disposable measurement cells contribute only negligible cost in many applications.

9.4.2 Cleaning

For on-line use, where measurements must be performed dozens to thousands of times per day, cleaning of a higher cost window or cell is a preferable approach. Chemical and/or mechanical cleaning mechanisms can of course be added to remove deposited films, either manually or automatically. In all cases this is expensive to perform, or reduces system reliability, risks more damage than help, or requires stocks of consumables, but it may still be useful.

Cleaning can be done in many obvious ways, including with water jets, air-jets, brushes, sponges, and ultrasonic cavitation scouring. Where hard windows of sapphire, diamond or diamond-like carbon are used, surprisingly tough cleaning methods can be used, including steel-blade scrapers, wire brushes, and abrasive stones. An interesting technique is the use of swirling grit particles and coated plastic balls, typically kept in motion by high sample flow rates (Fig. 9.9). These are frequently used to scour the electrodes of electrochemical sensors, but the idea is equally attractive for optical surfaces. In many cases these can be left in the measurement cell to sand-blast the windows during sample filling, falling out of the optical path by settlement or magnetic attrac-

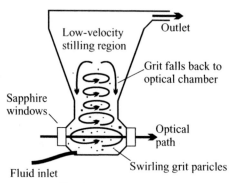

Figure 9.9 Swirling grit particles or neutral-buoyancy plastic balls can be used to scour the cuvette windows to remove contamination, and yet be retained with little loss.

tion during a measurement phase with stopped flow. If ferromagnetic particles are used, superior flow control may be possible using magnetic field drive from external coils.

Electrochemical cleaning can also be very effective, by generating reactive species such as OH-radicals via electrolysis, or by changing pH over a wide range in a region close to the sensitive windows. Simple heating can even dislodge some kinds of contaminants. Last, chemical cleaning using liquid- or gas-phase acid or alkaline chemicals and surfactants can be effective, especially if the chemical can be warmed. As with disposables, the replenishment of consumable cleaners brings problems and expense. Ideally sufficient stock is maintained at the instrument to last between periodic maintenance visits. Windows coated in toxic substances can supress biogrowth, although sample disposal may become a problem. Paints containing certain crystalline forms of titania (TiO_2) have been used to generate chemical radicals under UV illumination. These can be used for *in situ* cleaning.

9.4.3 Dissolving windows

Another approach to fouling removal is to use windows that *self-clean* via a slow dissolution process. Inorganic salts and special glasses have been used for this, usually bonded to an insoluble conventional window to avoid complete breakthrough if not replaced in time (Carr-Brion, 1996). They must still be considered to be consumables, which will contribute to running costs, and the surface quality of the window will be worse than a conventional (new) window. For systems that can tolerate some imperfection in the optical surfaces such as the integrating sphere described below self-cleaning is an elegant approach.

9.4.4 Automation

Automated systems also exist which combine cleaning with replacement. By using a pair of windows, one of which is in operation while the second is being cleaned, high data availability can be guaranteed. Changeover can be arranged to be automatic. The use of two windows means that more time can be spent cleaning, for example with long soaking in aggressive cleaners or with ultrasonic agitation. Automatically disposable windows are also possible. For example, tapes of plastic films can be moved continuously or in steps into the measurement cell, allowing periodic change of fouled windows. Last, we can try to avoid windows completely by operating the measurements windowless.

9.5 Windowless Measurements

9.5.1 Through-surface measurements

The highest stability, reliability, and independence of fouling contamination is likely when we can completely avoid touching the sample. This is easily possible in certain scattering-geometry measurements. Figure 9.10 shows a laser turbidimeter designed to measure through the free surface of a process

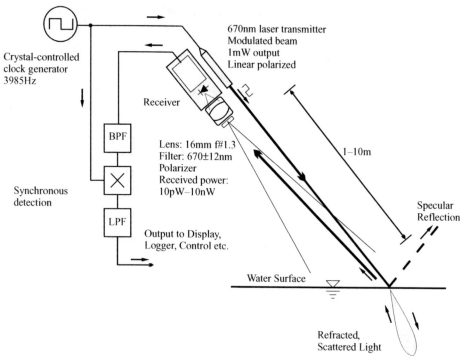

Figure 9.10 Laser turbidimeter uses the free surface to measure liquid scatter without contact or wetted windows.

stream. This is particularly convenient in the water industry, where open channels are used for transporting water, and in environmental sources such as rivers and lakes. The systems works as follows.

A small modulated diode laser projects a 1 mW, 670 nm collimated beam 3 to 10 m to the water surface. Apart from Fresnel reflection losses (≈2.5 percent), most of the beam energy refracts into the liquid, where scattering takes place at a distribution of particles of different sizes. A small fraction (typically 10^{-5}) of the incident light retraces its path to a collection lens focussed onto a single photo-diode receiver. For turbidities up to at least 1000 NTU the scattered intensity increases approximately linearly with turbidity. As the source and receiver are laterally offset from one another, but arranged with parallel axes, the laser beam and the acceptance cone of the receiver only overlap beyond a certain distance. With a 1-mm-diameter photodiode and a 16 mm focal length collection lens the acceptance cone is about 3.5° wide. Based on the principles developed earlier in the book, and noting that this instrument will need to function in brightly illuminated industrial halls and also in the open air with intense sunlight, the source is modulated at 4 kHz and synchronously detected. The received light level is only a few nanowatts, so a high value (≈100 MΩ) transimpedance is chosen. Nevertheless, with some care taken with component selection, amplifier gain-bandwidth product and transimpedance resistor parasitic capacitance compensation it is straightforward to achieve adequate detection bandwidth, over 8 kHz.

There are several drawbacks to performing the measurement in situ in a flowing process stream. First we have only limited control over the sample. Surface bubbles and floating objects will dominate the received turbidity signal, and must be removed hydraulically or via their distinctive temporal signals. The floating ring-tubes used by ornamental fish shops to improve our view of Koi do a good job here too to still the water surface and deflect bubbles.

Specular reflections of the laser source from the water surface will be much more intense than the diffusely scattered light, and will give large errors or even overload, which restricts operation to off-normal axis. If significant surface ripples are present, the range of reflection angles must be considered and a large-enough incidence angle provided. If the laser is polarized and the scattering process depolarizes, a crossed polarizer at the receiver can be used to reduce the specular reflections.

The liquid sample must also be "infinitely deep;" if the incident light propagates significantly through the sample to the channel wall or river bed, erroneous results will occur. This typically restricts application to moderately scattering process, environmental, and wastewater samples. However, these are just the media that benefit most from reduced contamination rate. An improvement is to offset the source and received beams to put the sensed liquid volume well under the surface, suppressing surface reflections and avoiding seeing the back-wall reflection. Finally, the open-beam character of the system brings problems with ambient light, in particular sunlight reflected from the water surface. By operating the reflectometer at Brewster's angle and using a crossed polarizer to attenuate the reflection, good suppression of sunlight is possible. This will generally be backed up by the electronic suppression techniques of Chap. 7.

The instrument is most convenient because of its ease of application. It is not necessary to provide a sample pumped through pipes, which is expensive and gives problems of pipe silting and abstraction blockage. It can be set up very quickly by mounting on available fittings, hand-rails, and furniture, and is useful even in temporary and covert installations. Last, the small beam means that measurements can be made where little access is available, for example through walkway grilles and observation apertures, and even in overflows, weirs, and pipe outlets. In the right applications, the technique can give turbidity measurements with a far lower cost of ownership than any conventional pumped-sample or immersed in-line turbidimeter.

Free liquid surfaces are so attractive for scattering-geometry measurements such as turbidity, fluorescence and Raman detection, that it is useful to make your own. If low fouling is the only benefit sought, much smaller separations than 10 m are adequate, and commercial turbidimeters are available from Hach and others with a free-surface provided by a pumped overflow in an on-line instrument (Fig. 9.11). Surface separations of the order of 50 mm are typical. High repeatability is possible in a well-maintained instrument. Free-

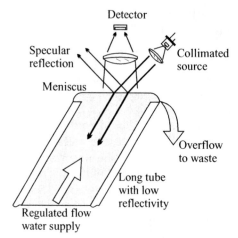

Figure 9.11 A free liquid surface can be generated within an instrument using a well-controlled flow and overflow system.

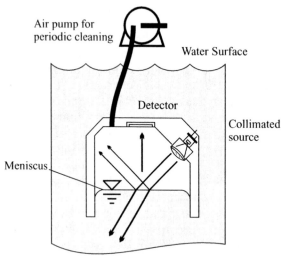

Figure 9.12 Free surfaces within the bulk of a liquid sample are possible using an inverted-cup. Occasional pressurization with a surface air-pump will clean away accumulated debris.

surface measurements can also be conveniently performed by forcing an inverted cup below the sample surface (Fig. 9.12). The hydrostatic pressure compresses the trapped air volume depending on the depth (1 atmosphere = 10^5 Pa ≈ 10 m water depth). This generates a controlled free surface, which can be positioned in the bulk of a deep sample. If the meniscus surface becomes contaminated, it can even be renewed by forcing air through the cup from a surface

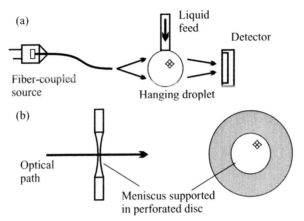

Figure 9.13 Small liquid samples may be supported by surface tension and probed in transmission without use of wetted windows, either using hanging droplets or a thin film.

pump. Both of these pumped and immersed free-surface geometries can be used for fluorescence detection and Raman spectroscopy, with similar nonfouling advantages. They are also useful for reliable long-term video-based observation of particles and flocs in industrial process streams.

9.5.2 Transmission measurements and the bubble interface

The nonfouling characteristic of the liquid-air interface is equally attractive for transmission-geometry measurements, although it is less obvious how to arrange this. A good starting point is the hanging droplet of Fig. 9.13a. By ejecting a small droplet from a capillary, an unbounded sample is formed which can be probed for optical absorption, fluorescence, etc. When small droplets (<2 mm diameter) are used they can be accurately spherical, and useful as a lens to focus light from a small source or optical fiber onto a detector. Bigger drops become more pendulous, leading to severe aberrations. Hence we are limited to rather small path lengths and poor sensitivity. Total internal reflection can be used to increase the path length, but this is difficult to control (Liu et al., 1995). Figure 9.13b shows another approach, more suitable for high viscosity liquids like oils and gelatinous film emulsions. A perforated disk is dipped into a bath of the sample and removed, leaving a meniscus supported in the hole. This can be used for windowless optical measurements free from gross fouling. As with the droplet, the path-length is rather short.

A more promising technique is the "holey-cup" of Fig. 9.14. Here we have drilled a 0.5 mm diameter hole in the base of a plastic cup, subsequently filled with water. As long as the hole is not too large, and the water head not too high, water will not flow through the hole, but will be retained by surface tension around the hole periphery. The 0.5 mm hole will support a 40 mm depth of clean

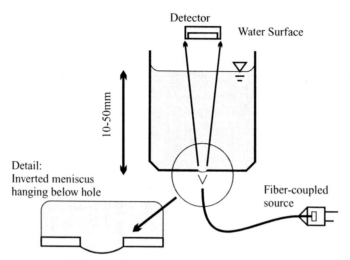

Detector

Water Surface

10-50mm

Detail:
Inverted meniscus
hanging below hole

Fiber-coupled
source

Figure 9.14 *Holey cup.* Surface tension is useful to support macroscopic liquid samples above a small hole, allowing noncontact optical transmission measurements through large sample thicknesses.

water without difficulty, although the water will bulge down into the hole to form a positive lens.

Again we have a macroscopic sample of liquid, bound on opposite sides by an air-water interface—just what we need for windowless transmission measurements. The small hole is certainly big enough to allow convenient transmission of light from LEDs (polished down to expose the bare semiconductor chip) or optical fibers. The meniscus curvature is even useful to collimate the diverging light. With the small size, surface tension forces dominate gravity, so that the meniscus is robust, and its shape is almost independent of orientation. This makes the technique useful for 90° scattering experiments also, when holes in the vertical container walls can be used.

The 40 mm path length is the same as in the longest cuvettes used with most spectrometers, and would be quite adequate for optical measurements in 96-, 384-well and other micro-titre cell arrays. Single holey-cups could be useful as very low cost disposable cuvettes for high volume measurements, even at wavelengths below 300 nm where expensive silica windows are typically needed. The hole makes a pretty good, smooth-surface window for use from the hard UV to the far IR.

In highly fouling media, even the holes will become blocked by settling debris, and eventually bridged by biofilms. Hence we need occasionally to blow the holes clean. This can be arranged, and further advantages gained, from the configuration of Fig. 9.15. Instead of leaving the underside of the supporting meniscus free to the atmosphere, we encapsulate it in a small-volume cavity with a conventional, heated window. In this way we can share support of the water column between surface tension at the hole periphery and pressure over the

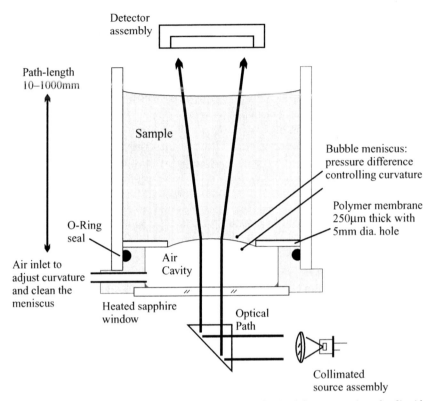

Figure 9.15 Better control of the holey bucket is obtained by supporting the liquid column primarily on a pressurized air cushion. The meniscus curvature may be adjusted using feedback control from the differential pressure. Plane menisci are possible, or we can make use of the optical power of a curved liquid surface for beam collimation.

whole meniscus area. Now the hole can grow to 6 or 10 mm diameter without great difficulty, allowing the use of large-scale sources.

By varying the pressure in the supporting cavity, the meniscus curvature can be varied from convex (positive optical power) to concave (negative optical power). With the correct pressure, the meniscus can also be adjusted to be accurately flat, suitable for collimated beam transmission. The pressure is of course a function of the water head, ambient temperature, surface tension, etc., so closed loop control is necessary to have a useful system. The figure shows a sensitive pressure sensor which can resolve a few microns change in water head, measuring the pressure difference Δp across the meniscus. As long as surface tension γ is constant, Δp defines the meniscus radius of curvature R identically:

$$\Delta p = \frac{2\gamma}{R} \tag{9.1}$$

An analog feedback loop automatically adjusts the pressure to a constant value, using a pressure actuator made up of two 40 mm diameter Mylar-cone loudspeakers arranged face to face as a small, fast pump. With a closed loop bandwidth of a few hertz, long term stable operation of the meniscus at a fixed pressure is possible.

The instrument we built (Johnson and Stäcker, 1998) was also fitted with a high flow air pump, which was actuated periodically to blast the measurement hole free of fouling. Depending on the degree of fouling, cleaning can be every few hours, or provide a new window for every measurement, analogous to the hanging mercury drop electrode so beloved of electro-chemists. The continuous loudspeaker-pump offers high speed and resolution but has a limited dynamic range. This can be improved by replacing it with a digital control system driving two pulsed solenoid valves. One was coupled to a pressurized cylinder at one bar overpressure, the other to a lightly evacuated cylinder. By rapidly pulsing the solenoid valves open and closed in less than 3 ms, small volumes of air could be introduced to or extracted from the support chamber. This approach gives limitless control, allowing automatic filling and emptying of the measurement chamber. A depth of 50–1000 mm was used and long-term noncontact measurements performed.

9.5.3   Falling streams: Transverse and axial

A well-known way to make windowless optical transmission measurements is the *falling stream*. Figure 9.16 shows a sample liquid container with a small hole in the base, through which water leaks out. The water thread that forms

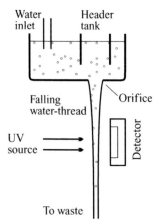

Figure   9.16 Falling   liquid streams and transverse transmission are used for noncontact transmission measurements of "difficult" liquids such as waste-water effluents.

under the hole is unstable, as surface tension eventually amplifies small diameter variations, causing the thread to narrow down into to a series of drops. A beautiful description of this process has been given by Boys (1959), along with drawings of uniform droplet streams, and alternating size streams. (Similar streams are sometimes seen along spider's webs. Spiders seem to be the experts of surface tension).

Although water threads are unstable, with reasonable care in design of the container's orifice, avoiding turbulence and carefully minimizing vibration of the container, surprisingly long threads can be formed in water and other liquids. The jets also often become electrostatically charged, so grounding of the apparatus is helpful. In the region above breakup, the thread takes on an eerily stable glass-like quality, eminently suitable for optical transmission and scattering measurements. The surface tension ensures that the higher spatial frequency components of surface ripples caused by orifice imperfections are rapidly smoothed out.

In transmission we have to live with the distorted cylindrical surface of the liquid. In principle this could be corrected with an external glass block, but in practice it is simpler to use a rectangular-section orifice, and achieve almost flat liquid surfaces for a few millimeters. Predistortion of the orifice might even move the optimum region lower down the thread to a more convenient location. Many products are available that use this falling stream approach, primarily in the ultraviolet where water absorption is highest, or with strongly absorbing samples such as waste-water and other effluents. This is just the application where some protection against fouling is required, so the technology is well suited.

Turbidity measurements on falling streams are also popular. When light is incident on an ensemble of scattering particles, the angular distribution of scattering efficiency is a function of the particle's shape, refractive index difference from its surrounding, and most importantly of its size. Particles much larger than the light wavelength predominantly scatter in a forward direction, while those much smaller than the wavelength typically scatter almost omnidirectionally. The latter case is convenient, as there is then little need to worry about the distortion caused by water surface shape. For best sensitivity the detector should be arranged to capture as large a solid angle of the scattered light as possible. Some applications, for example turbidity measurement of drinking water, use industrial standards that specify $90°$ scattering, which greatly degrades the method's potential sensitivity. The brewing business typically does allow small-angle scattering ($\approx 5°$), giving much higher sensitivity in the detection of low concentrations of yeast spores.

Multiple reflections occur between the water surfaces, which can be disturbing in some photometric and imaging applications. Forming antireflection coatings here to reduce the reflected power is likely to be a difficult task, but perhaps possible with oil films. Reflections are strongest for meridional rays, light travelling almost around the external surface of the fiber in *whispering gallery modes*, named after the famous acoustic effect heard in the dome gallery of Christopher Wren's St. Paul's Cathedral in London.

The falling stream can equally be used to improve the reliability of fluorescence and Raman detection, and a number of fluorescence-detection systems is available commercially. One system uses a low-pressure mercury vapor lamp emitting predominately at 254 nm in the ultraviolet to excite 270–300 nm fluorescence in a range of aromatic compounds such as benzene, toluene, ethylbenzene and xylene. This group (BTEX), with its origin in fuel-oils, is found widely in water-courses and environmental samples. It is hence a very useful marker to trace pollution effects, with their source in many thousands of leaking fuel-tanks.

Multiple reflections can also be put to good use to increase the sensitivity of turbidity, fluorescence and Raman instruments. The incidence angle for total internal reflection is 49° for an air-water interface, so the water thread can function as a high NA waveguide. Light scattered or emitted within the liquid thread will be guided to each end, where it can be detected. This can increase the effective detector capture angle, and hence the overall detection sensitivity. Figure 9.17 shows one configuration of fluorescence instrument, in which both the excitation light and the fluorescence are guided up the water thread to a detector. With a 200-mm long approximately one-millimeter wide thread a gain of six times was obtained above the unguided case.

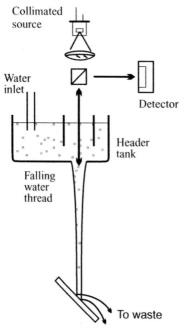

Figure 9.17 Axial propagation in a falling liquid thread is useful for non-contact scatter, fluorescence and Raman detection.

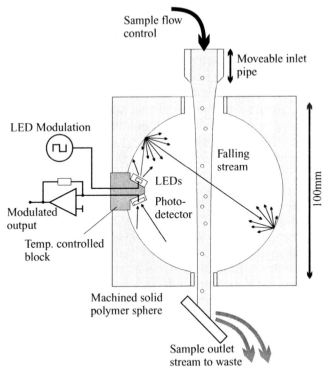

Figure 9.18 Combination of falling stream and the integrating sphere makes possible noncontact absorption measurement with reduced sensitivity to attenuation by scatter. Absorption depends only on the volume of liquid, not the shape. Individual drops work fine.

9.5.4 Integrating sphere plus falling stream

There is another falling stream system which is useful for highly fouling media. The integrating sphere of Fig. 9.18 is a hollow cavity made from or internally coated with a material with high diffuse reflectivity for the light wavelength of interest. The diffuse character means that after a few reflections, almost any input light distribution, for instance from an LED, will be transformed into a uniform optical field, with the power density being the same in all directions and at all positions throughout the cavity. The detected power-density at any point on the cavity surface can then be much greater than the input power divided by the surface area of the cavity. This assumes that there is no localized absorption within the cavity and that power losses through apertures, cracks, etc. are small.

These spheres are widely used to measure, or integrate, the total light output from sources such as incandescent bulbs and LEDs, whose total emitted power is to be determined. In practice they do not need to be even approximately spherical, but are used in whatever cubic, cylindrical or arbitrarily-shaped, more or less enclosed cavities, as the application requires. They are also used to determine the optical absorption of solid, liquid, and gas samples. Absorption can be measured by placing the sample within the cavity. With the introduction of an absorbing object or material, the average power-density in the cavity is reduced, and this reduction can therefore be measured to determine the object's total volume × absorption coefficient product. As the radiation is diffuse, the power-density reduction is very similar, independent of the shape or spatial distribution of the absorption, and independent of any loss-less scatter caused by particles in the fluid. Only the volume of absorbing material plays a role.

These characteristics are ideal for measuring the absorption of a falling stream. The shape of the stream plays little role; it is only necessary to control the flow-rate to maintain a given volume of material in the sphere. By using a diffuse light technique in this way we measure absorption independent of particle scatter, with sensitivity increased by the multiple path absorption, and without the inevitable fouling of a material flow-cell. Light-loss through the fluid entrance and exit holes reduces the optical gain due to the high surface reflectivity, but nevertheless good performance can be obtained with very "difficult" industrial process media.

9.6 Summary

Fouling is often considered to be an uninteresting annoyance to be solved with some bolt-on, low-tech cleaning contraption, and because of this it gets little attention in university research labs and publications. This is a pity. It is my feeling that solutions to fouling, through optoelectronic compensation, elegant cleaning techniques, new materials or total avoidance of contact with the sample is *the* most important aspect of many industrial measurements. Most on-line instruments are maintained far less often than the designer believes or expects, so it is important to design enough cleverness and innovation into the instruments to provide reliability and repeatability, even over a lifetime of misuse.

Measurand Modulation

10.1 Introduction

In earlier chapters we investigated techniques for optimizing the detection sensitivity of weak optical signals and for achieving high repeatability of measurements of relatively strong signals. The primary technical difficulties addressed were, respectively, noise and stability, which we treated as operating independently. Here we will look at an important measurement situation where the two errors are strongly coupled. This concerns the performance of an instrument in detecting very small variations in its measurand, that is, the instrumental characteristic of *limit-of-detection* or LOD.

Consider first the design of a high-sensitivity on-line scattered light measurement system, such as a turbidimeter used for analysis of drinking water (Fig. 10.1*a*). A high-power modulated light source is projected through the sample to a beam dump, while a high sensitivity receiver collects the scattered light, synchronously detects the source modulation, and displays the turbidity reading V_m. During commissioning the water is independently verified to be particle-free, so the relative light scattering efficiency δ and photocurrent I_p should be almost zero. As the signal is so weak, repeated readings of the instrument's display V_m are made and shown to be distributed around zero, as expected. This is represented by the fuzziness of the measurement result at the calibration shown as *Day 1*.

The next day V_m is found to have increased by a small amount ΔV to just above the first measurement value, causing us to ask "has the turbidity actually increased?" Again, repeating the measurement a few times gives the expected variation in results, which suggests that it wasn't a one-off error. The smallest change in signal ΔV that can be detected, and hence the LOD of the turbidity or other physical measurand, is clearly a function of the noise level of the detected signal. To be confident that the intensity has changed, for example between the Day 1 and Day 2 measurements, we require ΔV to be several times greater than the standard deviation of the noise, the measurement uncertainty.

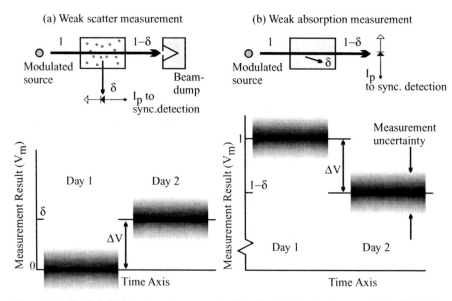

Figure 10.1 Dark-field detection, for example of low levels of optical scatter (*a*), is limited by noise and spurious scatter from the sample cell. Weak absorption measurements (*b*) are additionally limited by changes in source intensity and detection scale-factor.

However, the noise band given in Fig. 10.1*a* only shows the *high frequency* noise, that which causes variations during the period of the repeated measurements. In reality, the change ΔV may be due to other types of errors, for example baseline drift. If we perform a measurement without source modulation, errors will result from ingress of ambient light to the detector and from voltage offset drifts in the DC-coupled detection electronics. However, we have shown that these errors can be minimized using light source modulation and synchronous detection techniques. The only remaining sources of baseline drift are voltage offsets in the postdemodulation DC amplifiers, and these too can usually be kept to insignificant levels which do not limit the LOD defined by ΔV.

This leaves the errors of high-frequency noise seen during the measurement. The LOD can be improved, essentially indefinitely, by increasing the source power, increasing detection gain, and reducing the detection bandwidth, albeit at the expense of response time. In this way, very low turbidity levels can be and are routinely detected; with 100 mW of incident light and a 1 pW sensitivity receiver, we have a relative sensitivity of 10^{-11} of the transmitted power, or in practice the limit of the perfection of the beam-dump and suppression of stray reflections. This is a "dark-field" measurement; when there is no measurand property, turbidity in this case, there should be no detected light. Note that variations in source power, detector sensitivity and channel gain do not severely affect the LOD. They just affect the accuracy with which the LOD is measured.

For example, if the source intensity drops by 10 percent, this just gives a 10 percent reduction in LOD.

Figure 10.1*b* shows a quite different situation in which a weak, slowly changing optical absorption is being measured in transmission, and the available path length and absorption strength are such that the relative reduction in transmitted power δ through absorption is very small. Examples of such systems include trace chemical detection at the limit of an instrument's performance, monitoring the growth of bacterial colonies by absorption, and long-term environmental monitoring. These situations, with small and slow absorption changes, represent some of the most difficult examples of conventional optical measurement.

As before, a high power modulated optical source is available, almost all the power is transmitted to the detector, and the LOD is determined by the smallest change in relative power transmission δ that can be detected. For best performance, at commissioning the displayed reading is set up as 100 percent of full scale, and repeated measurements are shown to be distributed randomly about 100 percent. The next day V_m has decreased by a small amount ΔV to just below full-scale, so we ask whether the absorption has actually increased. What determines the smallest change in V_m that we can confidently say is due to absorption? As in Fig. 10.1*a* the standard deviation of high frequency measurement noise will limit the LOD, but this noise is likely to be small because of the high levels of received power and hence ease of achieving shot-noise limited performance. We will still be affected by baseline drift errors, but these will be minimized using source modulation and synchronous detection. Unlike the case of Fig. 10.1*a*, however, the LOD is additionally limited by *gain changes* in the measurement channel. If the gain changes by 10 percent, then $(1 - \delta)$ and V_m also change by 10 percent of full-scale, and this is the smallest change that can be confidently detected. This will usually limit the LOD to a value far greater than that defined by the S/N. As we are already operating with a full-scale signal, we do not have the opportunity to improve LOD by increasing the gain and reducing the bandwidth. Similarly, increasing source power and resetting the detector for full-scale doesn't help the LOD. Limitations on LOD due to errors caused by gain instability are likely to be many orders of magnitude greater than limits due to measurement noise. This is a fundamental problem of such a *bright field* measurement. If in Fig. 10.1*a* the relative signal due to δ, $\Delta V = 1$ percent and there is a 10 percent gain decrease, then ΔV changes by 10 percent to 0.9 percent of full-scale, a 10 percent error. If in Fig. 10.1*b* $\Delta V = 1$ percent and there is a 10 percent gain decrease, ΔV becomes 10x larger in value, a 1000 percent error.

10.2 Path-length Modulation

Let's suppose we want to detect a 0.01 percent change in transmission of a liquid sample. This would require a temperature stability of the order of 4 ppm/°C if a 25°C instrument temperature change is foreseen. If the source is

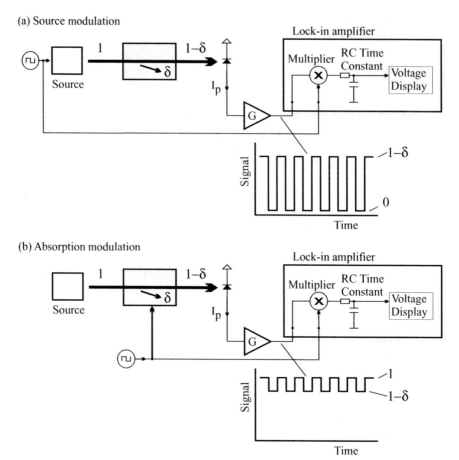

Figure 10.2 Source modulation (*a*) does little to improve the weak absorbance measurement, as the minimum detectable transmission change is defined by the source/detection stability. If the absorbance itself can be modulated (*b*), much smaller transmission changes can be detected.

an LED, we might need to stabilize its temperature to better than 1 mK to achieve this. While not impossible, this is tricky in an instrument for general laboratory or on-line use. This level of stability would be difficult to achieve with a conventional optical transmission measurement over a long period. A solution is to exchange source modulation for *measurand modulation*.

Figure 10.2 shows schematic optical transmission measurements. Our usual source-modulation scheme (*a*) gives a signal proportional to $1 - \delta$ pk-pk. The source modulation has removed baseline errors, but the signal is still proportional to channel gain, and in particular to the source intensity, gravely restricting the LOD. If we want to detect a signal change of 0.01 percent, we require a total stability at least as good as that. Even with all the stability and temperature coefficient issues of Chap. 8 addressed, this is a difficult and expensive requirement.

Figure 10.2*b* shows a scheme in which the absorption coefficient is somehow 100 percent modulated. Synchronous detection can still be used, although the detected signal is now much smaller (in this example about 10^4 times smaller). However, by making this change we have removed the requirement for unrealistic levels of source and gain stability. Measurand modulation greatly eases the requirements on scale-factor stability. If the gain changes by 1 percent, this only affects the measured peak-to-peak amplitude of the small modulation signal; the change in static intensity is not seen. Hence, for 0.01 percent resolution of the absorption, we only require that the noise of the transmitted signal is adequately small, not that the gain is stable to 0.01 percent. This is an altogether more realistic requirement. Now we will look at some of the means to modulate the measurand parameter of interest.

For instance we might vary the length of the transmission cell (Fig. 10.3), recording the change in transmitted intensity at minimum and maximum cell lengths, or even continuously as a function of sample length. Some care will be required in collimating the beam, in translating the cuvette window to avoid beam wobble and in choosing beam and detector sizes to avoid even small changes in capture efficiency. We cannot escape the requirement to reduce modulation-induced changes in detected intensity caused by these parasitic effects. Synchronous intensity changes caused by such errors have to be reduced to the full required relative precision, that is, to well below 0.01 percent. Nevertheless, the system intrinsically downgrades slow source power and detector sensitivity variations, especially caused by temperature changes, to signal scale-factor errors. Window contamination is also compensated.

The approach has been used by Horiba for in-situ absorption measurements in wastewater. In an elegant design, a fixed source and a detector are immersed in the fluid to be measured (Fig. 10.4). Eccentrically-mounted cylindrical silica tubes then rotate together, varying the sample path-length L over a large fractional change. The system suffers from a use of different window surfaces during different parts of the measurement cycle, which opens up additional

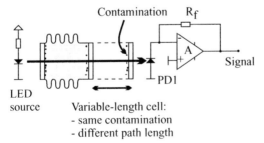

Figure 10.3 Variable cell-length gives a modulated intensity, with slow changes due to window contamination compensated.

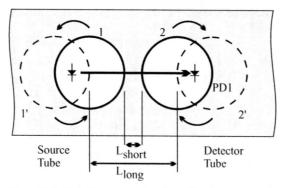

Figure 10.4 Rotating eccentric tubes allow in situ sample path-length modulation.

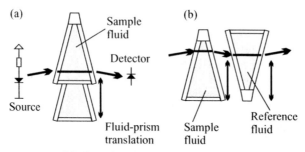

Figure 10.5 Moving prismatic sample containers (*a*) modulate the absorption. Dual cells allow pure sample referencing and a parallel optical path.

errors due to the synchronous intensity changes of nonuniform contamination on the transparent cylinders. These are expected to be small in comparison with the absolute intensity changes, and good performance is claimed.

Translated prismatic sample cells (Fig. 10.5) can alternatively be used where a linear translation is simpler to arrange than a rotation. The movable plane window cell (Fig. 10.5*a*) can either contain the sample fluid, or be evacuated or gas-filled and immersed in the sample solution. Dual cell systems (Fig. 10.5*b*) allow the use of a reference fluid related to the sample itself. For example, to detect small absorbances due to organic chemicals in water, a chemical-free water sample could be used as the reference. This has an advantage where we need to detect a small additional absorption in an already absorbing benign matrix. It is the situation in much of the optical testing of drinking water. The closer the match between the two liquids is, the easier it is to detect very small changes. If necessary the two prisms should be thermally coupled.

10.3 Concentration Modulation

In the majority of cases we have no control over the absorption strength of the unknown sample, and can only measure more or less sample as in Figs. 10.3 to 10.5. However, in the case of gas-phase samples, absorption can be modified through pressure variation, and this can be a very effective technique to improve sensitivity. The additional difficulty with gas sensing in the infrared is that the *signature* of absorption lines is very complicated, but the lines themselves are narrow. Hence if a broadband source is used to cover the whole width of the signature, even if the peak gas absorption strength is large, the reduction in integrated intensity received on a single detector will be tiny. Laser spectroscopy gets around this by using a narrow-band source which can be accurately tuned to an individual absorption line. If the line absorption is strong, intensity modulation as the laser is tuned across the line will also be significant. Pressure modulation gets around this by modulating the strength of all absorption lines simultaneously.

The conceptually straightforward approach uses a reciprocating piston to compress the unknown sample (Fig. 10.6). With suitable design of the piston and transmission cuvette, compression ratios of 10 to 100:1 can be achieved, and with it a similar variation in the absorption. Such systems for *pressure-modulation spectroscopy* have been designed for sensitive gas sensing. The mechanical complexity and perceived relative unreliability of conventional piston systems can be overcome using acoustic resonant cells (Fig. 10.6b). These are just "organ pipes" with the optical path arranged at an antinode of the standing pressure wave. Excitation can be via piezoelectric or electromagnetic actuators (loudspeakers), making use of the quality-factor (Q) of the resonance to increase the magnitude of pressure variation for a fixed input power. It is a pity that the incompressibility of liquids denies us this attractive modulation technique.

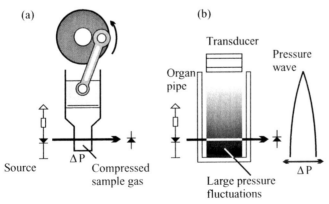

Figure 10.6 Pressure modulation of gas absorption using a reciprocating piston (*a*) and a resonant "organ-pipe" (*b*).

Gas absorption modulation can be combined with another related technique called nondispersive IR detection. Rather than using a broadband source, only a small fraction of which is absorbed by the gas's signature, it would be preferable to have a source spectrum made up of many lines which perfectly match the absorption spectrum. This is hard to arrange, but it is straightforward to produce a spectrum which is lacking the absorption lines. All we need to do is pass the broadband light through a cell of the target gas. As the absorbed wavelengths are missing, this coded light source should be only faintly absorbed by the sample gas. Modulation of the unknown sample should not produce a modulated intensity with the coded light, but a small modulation with the broadband light. These nondispersive techniques were developed in the 1930s, and a number of products using them is available, usually for detection of industrially important gases such as CO_2, CO, CH_4, SO_2, NO and H_2O. Originally detection was via gas expansion and movement of a pneumatic membrane. Different geometries are possible, including parallel comparisons of sample and reference gases, serial transmission, and high-speed switching of sample and reference gases.

10.4   Flow-injection Analysis

During the 1960s and 1970s techniques for automated chemical analysis were developed to handle large numbers of tests performed on blood and urine samples. These used continuous liquid flows in small ($\approx$mm diameter) tubing, driven by a multichannel peristaltic pump. Further elements were added to allow the mixing and reaction of solvents, pH buffers, colorimetric indicators and liquid standards, the injection of known volumes of the sample liquid under investigation, residence periods provided by longer transport delay tubes, and in-pipe detection systems, for example by optical absorbance. A typical system is shown in Fig. 10.7. Originally samples were kept separate from each other during their passage through the tubing by interspersing small bubbles of air or other gas (Fig. 10.8). These *segmented-flows* serve to reduce hydraulic dispersion and intermixing, and to scour the tubing wall to reduce build-up of contamination. Later it was found that in small-bore tubing axial mixing could be kept to negligible levels even without the use of gas segmentation; many such *flow-injection analysis* or FIA systems currently do not segment the sample between gas bubbles.

Although the primary drivers in development of FIA systems were undoubtedly sample throughput and cost-per-test, the idea of mechanically moving samples past a detector also provides an effective measurand-modulation scheme. By programming the systems to perform repeated measurements on chemical standards, interspersed by the samples, slow changes in the fouling, optoelectronic system baseline drift and even absolute scale-factor can be removed. Several configurations are possible depending on what reference fluid is used (Table 10.1).

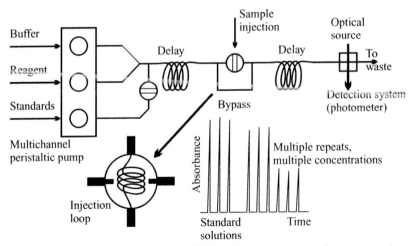

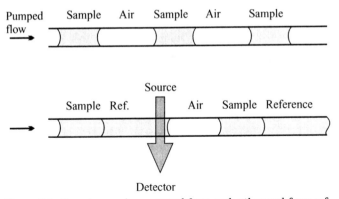

Figure 10.7 Flow-injection analysis (FIA) modulates the sample concentration, making detection, even with short path-length optical detectors, more effective.

Pumped flow → Sample   Air   Sample   Air   Sample

Source

Sample   Ref.   Air   Sample   Reference

→

Detector

Figure 10.8 Samples can be separated from each other and from reference liquids using bubbles ("segmented flows"), although these are not needed if the pipe diameters and flow rates are chosen correctly.

TABLE 10.1 Referencing Choices for Water Absorption Measurement

Primary Measurement	Reference Measurement
1. Sample water	Air/Gas bubble
2. Sample water	Pure water
3. Sample water	Purified sample water
4. Sample water plus indicator chemical	Sample water
5. Sample water buffered at acid pH	Sample water buffered at alkaline pH

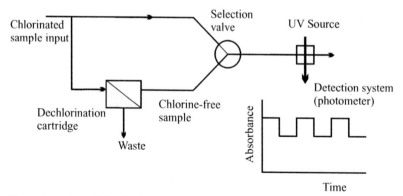

Figure 10.9 A useful technique is to derive the absorbance-free reference liquid from the sample itself, for example via dechlorination.

For example, in the detection of trace absorption changes in a more or less pure water sample, we could arrange for sequential measurement of an air reference, a sample of pure water, and the unknown aqueous sample. By measuring at the same point in the tubing with a single source-detector system, tubing wall contamination errors can be substantially compensated, while the small differences between sample and solvent can provide a better measure of the small absorption increases.

In the measurement of very weak absorptions, it is preferable to use as reference a fluid as similar as possible to the sample itself, so that the detection system can be designed for high amplification of the small differences. This promotes configuration 3 of Table 10.1, in which the reference is obtained from the sample stream itself, except purified of the component which is being sought. A typical example might be an optical chlorine disinfectant measurement system, where dechlorination is achieved by passing the water through an activated carbon filter (Fig. 10.9). In practice it is very difficult to remove only the target species without also removing some of the background matrix absorption.

Chemical indicators can be used for high sensitivity determinations of a wide variety of chemical species. The reactions are often selective, and the intense colors of some of the reagents give much higher sensitivity than when species absorption is directly measured. Nevertheless, compensation for the absorption of the sample stream is necessary for the best performance, and FIA is a convenient way to automate this. Verma et al. (1992) describe a system for residual chlorine detection in which the reaction product is a type of azo dye which absorbs at 532 nm. This wavelength is much easier to detect, and water absorption is much lower there than at the chorine absorption wavelengths of 234 nm and 292 nm. The FIA system used performed measurements at a rate of 110 hour^{-1}.

10.5   Electromodulation Analysis

In a limited number of situations the absorption strength of an unknown liquid sample can really be modified, and this forms the basis of an attractive detection scheme. Pesticides and herbicides that find their way into drinking water supplies can in principle be quantified using the strong ultraviolet absorption of these organic chemicals. However, the very low concentrations which are of interest (≈ppb) give only very weak absorption, in a drinking water matrix with its own, much stronger absorptions due to unimportant dissolved inorganic and vegetable matter, whose magnitudes also change from hour to hour. We need to make the reference and sample measurements with as little delay as possible, and with as little change to the mechanical configuration as possible. In this case it is a big help to make a difference measurement using the sample itself as reference, as shown in Fig. 10.10. This is an example of a *stopped-flow* system. First a UV/visible spectral measurement is made of the water sample as received, for example pumped in from a fast-flowing sample line but stationary during the spectral measurement. Then an intense flash of light from a xenon flash tube illuminates the sample that is about to be measured. Several herbicides can be rapidly photolytically degraded in this way, while leaving the remainder of the water matrix absorbance spectrum intact. We measure the spectrum once again, and then calculate the difference in software before pumping in a new sample. This difference spectrum is characteristic of the herbicide present, with the bulk of the time-variable matrix absorption ignored. In some cases, for example the 292 nm absorption feature of alkaline chlorine disinfectants (HOCl) in drinking water, the charac-

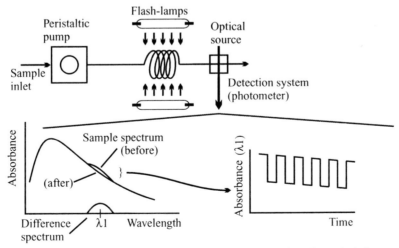

Figure 10.10   In-situ flash photolysis is a technique to modulate the optical absorption of a photo-active species in a complex nonactive matrix. Subtraction of the before/after spectra reduces the effects of absorbance drifts due to the matrix.

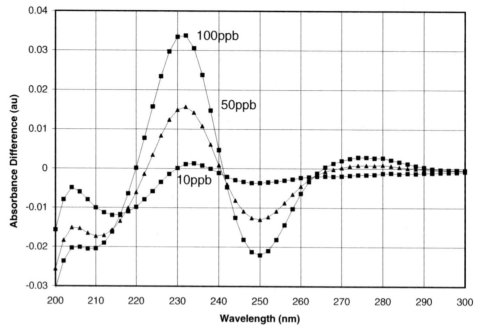

Figure 10.11 Difference spectra of 10, 50, 100ppb atrazine in water formed by in-situ flash photolysis.

teristic absorption is simply bleached out by the photolysis reactions. This is just what happens to swimming pool water in sunlight. In other cases, including many of the herbicides, the absorption features of the herbicides are not just *removed*, which would lead to a positive absorbance value for the before-after difference, but modified by the photolytic action, giving both positive and negative spectral features. This can aid determination of the types of contaminants present. Figure 10.11 shows difference spectra of three concentrations (10, 50, 100ppb) of Atrazine in water, modified by a UV/visible flash pulse.

This photolytic modulation is just one example of electrochemical modulation schemes which are useful for modulating the intrinsic absorption of active species such as benzene, toluene, ethyl-benzene, xylene ("BTEX"), phenol, aqueous chlorine (HOCl/OCl⁻), and many common indicator dyes. Other modulation techniques include thermal degradation, pH changes generated at electrodes, and the use of high voltage corona discharges. Figure 10.12 shows an electrochemical reactor in which a *corona-wind* of electrons and charged species is injected into the surface of a contaminated water sample. The liquid is grounded via an electrical connection in the silica cuvette base. The corona-wind is generated by a high tension power supply providing current of a few microamperes. The injected species can modify contaminant molecules as does the photolytic illumination. Figure 10.13 gives results for a 0.0, 0.5, 1, 2ppm

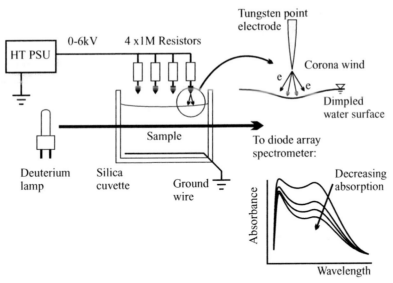

Figure 10.12 Corona-wind modulation spectroscopy. Charged particle injection from a high-voltage corona can be used to degrade and modify a wide variety of organic samples. The system is easily installed within the measurement spectrometer.

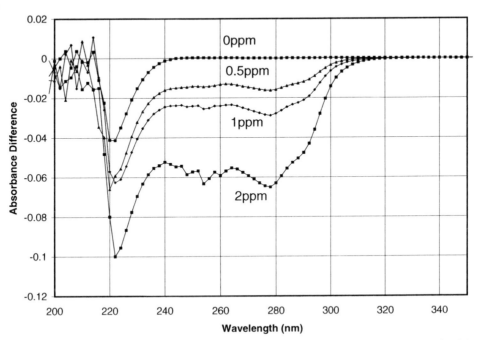

Figure 10.13 Corona-wind modulation spectroscopy of benzene in water at concentrations of 0, 0.5, 1, 2 ppm.

solution of benzene in drinking water. The difference spectrum is all negative here, as the absorption is increased by this treatment. The uppermost curve is for pure water. Note that there is some absorption change even here, starting below 240 nm wavelength. It is probably caused by modification of nitrate ions present in large concentrations in this water. However in the 240 to 300 nm region the effects are small. Away from the modulated absorption features, the difference is also small as the corona discharge has little effect on the background matrix. See Melbourne (1996) and the author's papers.

The best feature of electrochemical measurand modulation is the high stability of the measurement configuration between operations. The corona and photolytic modification steps are performed within the spectrometer, without physically replacing samples between measurements, and without mechanical movement. This helps to reduce other synchronous changes in intensity. All the other techniques, such as path length variation and FIA liquid movement introduce the risk of significant changes in optical transmission even with no sample present. As we saw, the LOD is now limited by remaining synchronous parasitic variations, which should be minimized by stable geometries, repeatable optical path losses, and rapid sample/reference measurements.

10.6 Mechanical Scanning

Flow injection analysis uses physical motion of fluid samples in fine-bore tubing to provide, among other benefits, optical measurand modulation. The system is a "liquid-chopper." However, relatively large quantities of buffers, reagents and sample ($\approx$mL/min) are needed. As the available sample quantities decrease, it may be more convenient to capture μL samples on solid media such as filters, paper tapes or synthetic films, and to move the support media mechanically. Figure 10.14 shows a diffuse reflectance measurement system

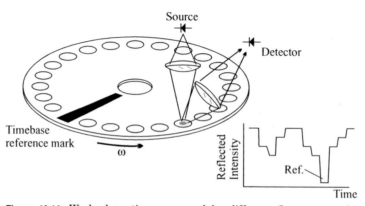

Figure 10.14 Weak absorptions measured by diffuse reflectance can be detected at higher resolution by mechanically modulating the samples, spotted onto a uniform filter paper.

configured for multiple small samples. A track around the perimeter of the 50 mm diameter paper disk can hold about 50 samples each of $1 \mu L$, either all different or multiple instances of the same sample. Two timing mark systems should be provided, one to signal the start of a revolution, the others to signal the angular position around the disk. They can be provided either on the support arbor or printed onto the paper itself. As the disk spins, each sample is read out and the area under the absorption signal is digitized in time. Similar systems have been built based on chemical analyses done in multilayer films in tape form. With a single sample divided into multiple patches and spun at high speed, we effectively have a diffuse reflectance modulated by a periodic modulation of sample concentration. Large improvements in LOD are possible with this approach.

10.7 Remote Sensing: Fiber Refractometer Revisited

In the modulation techniques described so far, the instrument has been available, and sample modulation has been straightforward in principle. In the case of optical fiber sensors, however, the location where we would like to modulate may be in a remote or inhospitable place. We have already seen the difficulties of achieving even one-bit of resolution in a fiberoptic reflectometer. Due to fiber attenuation which can be highly variable from installation to installation and over time, even detecting the presence or absence of a switched mirror can be difficult. The signal of interest is, because of attenuation, a tiny fraction of the swamping intensity due to unterminated fiber ends, connectors, and material back-scatter. Modulation of the source, although convenient for intensity quantification and industrial interference removal, does not improve the S/N of the true signal. However, if we can perform the modulation at the distal end of the fiber, that is, of the light incident on the sensor-head, then large improvements are possible.

Figure 10.15 shows a fiber refractometer with a mechanical intensity modulator, formed from a stepper-motor driven chopper wheel. If this is battery-driven, or powered via a solar cell from ambient light or light guided in the fiber, and the power consumption can be kept of the order of $1 \mu W$ or less, then convenient and low cost-of-ownership fiber sensors can be constructed. Even micro-electro-mechanical-systems (MEMS) clockwork motors driven by mechanical strain energy may be attractive. With a low-reflection chopper wheel, the signal intensity will change periodically, with the intensity difference being that due to the sensor head reflection. The received signal will contain the disturbing reflections in the unterminated reference fiber arm ($9 \mu W$), other fiber arms and in-line connectors, with superimposed the small modulated intensity from the liquid under test (minimum 1.5 nW). By reducing the detection bandwidth we should have no difficulty in detecting the refractometer signal, with an accuracy determined by the intensity stability, detector tempco etc. Without the measurand modulation, detection of small changes in the 1.5 nW minimum

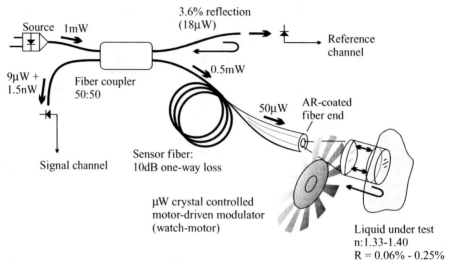

Source 1mW
3.6% reflection
(18μW)
Reference
channel
9μW + 1.5nW
Fiber coupler
50:50
0.5mW
50μW
AR-coated
fiber end
Signal channel
Sensor fiber:
10dB one-way loss
μW crystal controlled
motor-driven modulator
(watch-motor)
Liquid under test
n:1.33-1.40
R = 0.06% - 0.25%

Figure 10.15 Clockwork, battery, solar or fiber-powered modulators placed at the point of measurement are useful to aid the determination of weak intensity reflection changes, which would otherwise be swamped by fiber loss variations and spurious reflections from other fiber components.

reflected power on top of a variable $9\,\mu W$ signal looks very difficult. This approach could significantly improve detection of the small and slow intensity variations caused by the liquid under test. If in addition the switching frequency is very accurately known, then the reference-less synchronous detection schemes of Chap. 5 can be used for narrow-band detection. We can note that the specifications of the microwatt stepper motor chopper are pretty well exactly provided by a modern electronic analog watch, battery-operated for a life of many months.

Many other schemes for remote intensity modulation have been published over the years, using electrooptic devices driven by photo-cell electronics, optooptical modulators, polarization-coding, and wavelength coding. The principles of applying modulation at the point of sensing to avoid much of the problem of variable attenuation in the fiber leads are similar.

10.8   Discussion of the Choice of "Modulation Frequency"

In the discussion of Fig. 10.1 we suggested that fluctuations in channel gain caused by source intensity variations and component tempcos could really just be treated as another noise source. If the frequency-spectrum of that noise, including all drifts and other errors, really is *white* and Gaussian, we know that the longer we go on making measurements, the greater might be the absolute variations seen, but the mean and standard deviation of the distribution of values will be constant. Also, the longer we go on making measurements, the better we will define this mean. This just isn't how drift and temperature errors seem to work! For example, the precision of measurement of a shot-noise limited $100\,\mu A$ photocurrent is typically far worse than the noise calculation would suggest. Experience seems to say that the potential error depends in some way on the time-scale, for example on the time since last calibration. The longer we wait, the greater is the likely error. This is, after all, the basis of industry's reliance on periodic calibration. In order to understand the errors, we must estimate the measurement noise frequencies and related timescales, which depend on the type of measurement we are making.

The first is an *absolute* measurement, where we are given an unknown sample, and asked to measure the absorption. We just insert the sample and press "read." The instrument was calibrated three months ago, so the important characteristic is the variation that is likely to have taken place since then. This is given by the noise spectral density at the corresponding frequency, something like 0.33 month^{-1}. An alternative and much easier measurement is where we have a chance to zero the equipment with a *blank*. Much of laboratory spectroscopy is carried out in this way. First we measure a cuvette of deionized water and call the transmission unity. Then the real sample's transmission is measured, and the intensity ratio is computed. In this case the pertinent frequency is 1/(time between the blank and sample measurements) or perhaps 1 minute^{-1}. In the case of gas absorption modulation in the organ pipe of Fig. 10.6 the frequency could be $\approx$kHz.

For some instruments such as electrochemical sensors used to measure chlorine disinfectants in drinking water the *noise-and-drift* is quite different. Here the dominant problem is fouling, and the sensitivity of the electrochemical process essentially just degrades with time. If the sensor is used as input to a closed loop dosing control system the true concentration will be automatically adjusted upward to compensate for the reducing sensor sensitivity, until the electrodes are cleaned and the instrument recalibrated. The resultant "sawtooth" drift function is clearly quite unlike Gaussian white noise, and the calculation of the mean signal and standard deviation depends on the time-scale evaluated. Similar problems arise with optical transmission systems which do not include compensation for window fouling. They become gradually dirtier. White noise just doesn't describe these processes.

As another possibility, what if the noise-and-drift spectrum really has a $1/f$ character, as sketched in the upper part of Fig. 10.16. After all, we know that this is prevalent in active electronic components and many natural processes. Then the noise power density for frequencies corresponding to a year since calibration (1/year) would be much greater than those for blank and sample measurements separated by a minute. The longer we wait, the greater the likelihood of measuring with a larger error, and the longer we go on measuring, the greater becomes the standard deviation of the distribution. This seems to corresponds much better to practice. The time-series of the lower part of Fig. 10.16 was calculated by setting up a frequency distribution of random

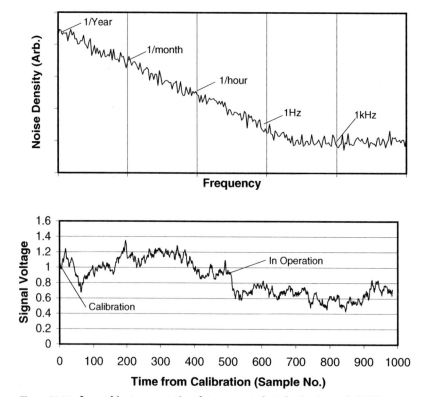

Figure 10.16 Logged instrumentation data suggest that the "noise-and-drift" power spectrum is not white, but increases at low frequencies, giving errors far greater than expected from the instrument's optical noise performance. The time-series was calculated using a $1/f$ power spectrum, and looks like real data. The greater the period between calibration and operational use, the greater the measurement variance.

Gaussian-distributed complex values with a $1/f^{\beta}$ power spectrum, here with $\beta = 1$, and then using the discrete FFT to transform to the time-domain. It looks similar to real measurement drift in an instrument free of ever-increasing fouling, with small variations seen on short time-scales and much larger variations over a longer period. The same approach can be used to investigate other frequency distributions. With $\beta = 0$ we have white noise, with $\beta = 2$ Brownian noise or a *random walk*. Of course we generally do not know the details of the noise-and-drift spectrum of our instrument at these very low frequencies. There may for instance be a break frequency above which the noise density is white, as in most semiconductors. There may be spot frequencies corresponding to diurnal variations in calibration. Nevertheless, if $1/f$ noise is a better description of reality than is white noise, it appears that the reduction in drift in going from the 1/year frequencies of annual calibration to 1/ms of the spinning diffuse reflectance equipment can be very large. Hence we should certainly try to modulate our measurand fast enough.

It would be a great help to understand the noise-and-drift spectrum down to 1/year frequencies to optimize recalibration regimes, and up to kilohertz frequencies to optimize measurand modulation. FIA systems, in-situ photolysis, and the spinning disk of Fig. 10.14 can all be considered as absorption modulators for liquid samples. We can change the modulation frequency in FIA through the segment volume and the flow-rate, and for the disk via spot-size and rotational velocity ω. To make best use of the modulation, we need to choose the modulation frequency appropriately. The considerations involved in the choice are the same as those considered in temporal modulation in a lock-in amplifier measurement: what is the frequency distribution of noise sources, and can we choose the modulation frequency to avoid the worst regions? For example, independent of the $1/f$ noise character, in the case of photolytic absorption modulation of a drinking water sample we might find that the background matrix changes significantly over a period of an hour due to quality changes. Hence we should certainly modulate with a frequency faster than 1/hour. If we can make one flash/measure measurement cycle per minute that will probably be fast enough. With the development of low cost MEMS the use of microfluidic devices, which can rapidly switch nanoliters (nL) volumes of sample, may open up new applications of high-speed FIA-type measurand modulation.

In the case of moving disk or tape reflectance measurements, we have a lot of control over modulation frequency through the rotational velocity. Just as with source modulation and synchronous detection it is probably a good idea to modulate at a frequency well above the level of industrial noise sources, say above 500 Hz. With a 50 mm diameter disk and ten sample spots per circumference this is possible with a rotational velocity above 3000 rpm (linear velocity >8 m/s). However, this may not be optimum in terms of spatial frequencies. As the disk rotates, even without any sample present, the detected intensity will not be perfectly constant. Due to bearing wobble and imperfect centration there is likely to be a detectable intensity variation at the fundamental frequency ω. If the filter paper has been calendered or otherwise formed between rollers it is likely to have uniaxial surface structure and different angular scattering characteristics along and perpendicular to the calender direction, giving a frequency component at 2ω. At very high spatial frequencies we will begin to see the individual paper fibers. Hence we have to address both *temporal* and *spatial* frequencies. A single 50 µL sample absorbed to cover one half of the filter disk is likely to suffer from the low spatial frequency reflectance variations. If it is divided into thousands of 100 µm spots fired on around the circumference from an ink-jet actuator fiber noise may provide the limit to LOD. We may find that optimum performance will be obtained by having 20 1-mm sized spots distributed around the circumferential track, 1/mm being well above the dominant manufacturing nonuniformity frequencies, but below paper-fiber noise frequencies. This choice is of course independent of the rotational velocity.

10.9 Summary

We have shown that the ultimate sensitivity of many optical instruments is restricted to a poor value not by optical-system measurement noise, but by low-frequency drifts in system gain. Source modulation does nothing to reduce these errors. They can be orders of magnitude greater than expected from the electronic noise of a typical shot-noise limited detection system, which suggests that the noise spectrum cannot be flat, and drift is not simply low-frequency Gaussian noise. If the noise spectral density increases at low frequencies, the longer we wait, the greater will be the error, and we can improve the ultimate sensitivity (LOD) by modulating the optical measurand at sufficiently high frequencies. The most promising techniques involve electronic modulation of the physical measurand, ideally performed within the instrument at the point of detection. However, to best use these systems, we will need a much better understanding of the noise and drift performance over a wide range of frequencies.

Multiple Channel Detection

11.1 Introduction

Until now we have restricted ourselves to single-channel optical transmission systems, and their optimization with low-noise design and narrow-band synchronous detection. In Chapter 8 a second, reference channel was added to reduce instability caused by source temperature fluctuations and aging. However, this still only allows the measurement of a single optical channel. For many kinds of optical instrumentation, the single channel measurement is too restricting, and we must consider how to configure multiple channels in a convenient way.

11.2 Chemical Analysis and Optical Filters

Consider an application from chemical analysis, in which the concentration of a liquid colorimetric reagent in an environmental water sample is to be determined from its optical absorption. This is essentially a realization of the two-wavelength schemes of Fig. 9.5. The background water is absorbing, and this absorption varies from day to day as rainfall modifies the source's flow and content. It is decided to measure at two (or more) wavelengths, chosen from an investigation of the reagent's absorption spectrum. Figure 11.1 shows a UV/visible absorption spectrum of a dilute solution of potassium permanganate. The primary absorption peak of this reagent lies around 540 nm. A readily available pure-green LED emitting at 525 nm is a reasonable match to the absorption feature. Additionally a 430 nm GaN LED is used to see predominantly the reagent's low absorption region. Both these wavelengths are conveniently addressed using readily available LEDs, and with the choice of devices it is often possible to find useful combinations. If not, color filters could be used, either in addition to the LEDs or with chopper-modulated incandescent sources.

The chemical concentration is calculated from the ratio of attenuations at the two wavelengths, or the difference in absorbance values, which hopefully

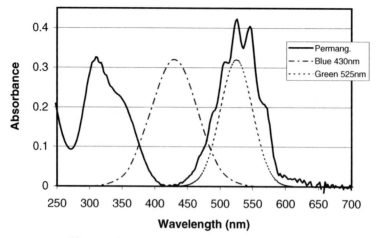

Figure 11.1 The concentration of chemicals with well-defined absorbing and loss-less regions can be determined with high resolution by measuring at two wavelengths on and off the absorption feature and subtracting the absorbance values.

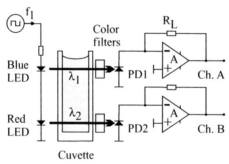

Figure 11.2 Multiple channels can be separated spatially and with matched color filters.

compensates for the bulk of any additional water matrix absorption. We will use source modulation and synchronous detection, and a brute-force approach with duplicated optoelectronics (Fig. 11.2) would do the job. The two source beams with different colors are arranged side by side in the cuvette holding the sample. Separation of the two channels is primarily spatial, although in order to reduce problems of the two beams seeing different parts of the liquid and different cuvette surfaces, we would like the beams to be as close together as possible. On the other hand, proximity will increase the risk of one source illuminating the wrong detector. Therefore we additionally isolate the channels using individual color filters in front of the detectors. These also serve to reduce the detection of ambient light. To aid in output-power tracking of the two LEDs, each is driven with the same current, modulated at the same frequency f_1 for quantification using two lock-in amplifiers.

While the brute-force approach will be effective, its color filters are expensive, especially if they must be custom sawn into small pieces or odd shapes. Further, the filters enforce separation between the two beams which means that the two channels never probe precisely the same sample. This can be very significant in on-line applications such as industrial manufacturing and water treatment processes, where inhomogeneities and even multiphase flows exist. It can also be a problem in laboratory analyses, due to poor mixing of reagents, settlement and nonuniform adsorption of materials onto the container walls. Improved performance is likely if we overlap the optical paths as closely as possible. This is easy to do using any of the beam-splitters of Chap. 8 (Fig. 11.3), but again this significantly increases both complexity and cost. On-line photometric instruments configured like this are widely available.

Instead of color filters we could use 90° aligned polarizers at source and detector, or polarization beam-splitters just as in Fig. 11.3. Even plastic film polarizers can achieve <0.1 percent crossed-state isolation, and crystal polarizers can better 0.001 percent. This is of course only useful if propagation through the scattering medium does not significantly depolarize the light. Polarization-coding is limited to just two channels.

A further error in these ratiometric measurements using separate detectors is the lack of perfect tracking of the sensitivity. They are likely to be at different temperatures, causing errors through their temperature coefficients. Good thermal coupling will help. Separate detectors also make severe demands on mechanical stability. Any vignetting or displacement of the beams which effects one photodiode more than the other will lead to errors. As it is so hard to ensure that these errors are adequately small, many designers try to get by with a single detector. The synchronous detection process makes possible a large number of useful configurations.

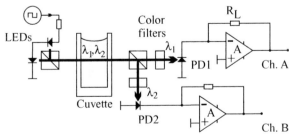

Figure 11.3 Best results are obtained if two or more beams can occupy the same region of the sample and cuvette windows.

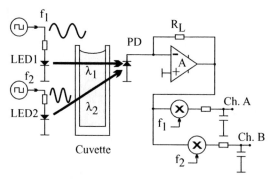

Figure 11.4 Frequency-coding of two or more measurement channels.

11.3 Two-Frequency Detection

Figure 11.4 shows a two-channel absorption measurement system where each wavelength channel source is modulated at a different frequency. A single detector and transimpedance amplifier are used, which helps to optimize the tracking of detection gain of the channels. The photodiode's temperature coefficient of responsivity is a function of wavelength, so temperature variations will still lead to some differential error, but in a low-tempco photodiode this differential error will hopefully be very small, and certainly smaller than if two separate photodiodes are used.

Path equality for the two wavelengths requires close juxtaposition of the two LEDs or even interspersing of multiple LEDs in a single cluster, and this is generally straightforward to do. It is true that we have lost the advantage of using the same current drive in both LEDs, but as the LED temperature coefficients are quite different anyway, we haven't lost much of an advantage. How should the modulation frequencies be chosen? On one hand it would be better to keep them bunched close together, in order to make best use of available bandwidth in the photo-receiver, and to match the dynamic performance of the two signals. On the other hand, if they are too close together, interchannel interference will result.

We saw in Chap. 6 that the transmission frequency passband of the synchronous detection process is centered on the modulation frequency f_{mod}, with a width proportional to the reciprocal of the postdemodulation low-pass filter time-constant. However, the passband cutoffs are not infinitely steep; typically they achieve only $-20\,\text{dB}$ or $-40\,\text{dB/decade}$, perhaps necessitating two decades of separation for good isolation. Hence with a $1\,\text{Hz}$ detection bandwidth, and as long as the modulation frequencies are stable to the same order, separation of the modulation frequencies by $100\,\text{Hz}$ should give negligible interference. If higher detection bandwidths are needed to track fast-changing measurands, the separation will need to be increased in proportion, which can become inconvenient.

However, this is not the only source of interchannel interference. We saw that square-wave demodulation produces harmonic responses at $3f_{mod}$, $5f_{mod}$, etc. If the detected signals are also not perfectly sinusoidal, they will contain energy at these harmonics which will be detected to some extent by the harmonic responses. It is therefore also necessary to avoid overlaps of these harmonics from different frequencies f_{mod}, matching to within the detection bandwidth. It is probably only worth worrying about the first few coincidences, for example $3 \times f_{1mod} \approx 5 \times f_{2mod}$, but if the application is critical or the performance marginal, all such coincidences must be investigated and their interactions measured.

11.4 Two-Phase Detection

11.4.1 Sine/cosine modulation and detection

For two-channel only systems it is possible to use instead a single modulation frequency f_{mod} with two different phases (Fig. 11.5). This can be arranged with a single modulation generator, providing analog sine/cosine waveforms to the sources, or more simply 90° phase shifted square-waves. After detection at a single photoreceiver, two synchronous detectors can be used to extract the modulation amplitudes. The detection process forms the product of the received waveform with either sine or cosine, and as these are orthogonal, one channel should not see the other channel wavelength. This is just the definition of orthogonality, that the integral of the product is zero.

Many conventional lock-in amplifiers already include the two multiplier/filter detection channels, and even an accurate 90° phase shifter for generating sine and cosine (quadrature) signals from a single reference input. Hence, with careful adjustment of the detection phase, this type of two-phase coding and detection can be performed in a single off-the-shelf unit. In practice, inter-channel orthogonality is not perfect, due to imperfection of the source drive waveforms, distortions, nonlinearities and group delays in the receiver, and

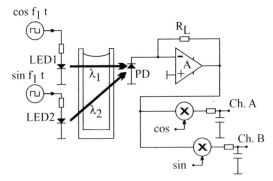

Figure 11.5 Phase-coding of two measurement channels.

imperfect phase adjustment. Nevertheless, the lock-in realization of this approach makes a highly didactic demonstration, which for moderate isolation is effective.

11.4.2 Multimode fiber refractometer revisited—again

In the design of a simple refractometer for remote liquid refractive index measurement shown in Chapter 8 we wanted to measure two signals, one a reference intensity representing the power coupled into the fiber system, the other a variable intensity reflected from the sensing fiber end. One problem was how to suppress the reflection from the monitor fiber arm to very low levels (at least −60 dB), such that it does not perturb the weak reflection from the sensing probe. One way to do this is to treat it as a two-channel measurement system, and to apply the two-phase detection technique.

If the light source of Fig. 11.6 is square-wave modulated at f_{mod}, the interference and signal intensities will be superposed on the signal photodiode, but due to group delays they are not necessarily aligned in time. So far we have only considered group-delays caused by RC-networks in the photodiode and receiver components. However, with the long transmission lengths typical of fiber

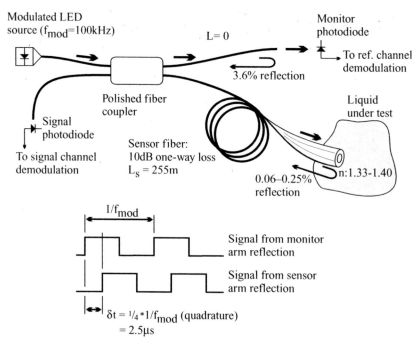

Figure 11.6 Phase-coding applied to a long fiber reflectometer. The main reflections from the sensor and from the reference arm can be arranged to be in quadrature through correct choice of the source modulation frequency.

sensors, there is also the propagation group delay to consider, even with relatively slow modulation. This is given by: delay = distance/velocity, where we use the group-velocity in the fiber (v_g), not the phase-velocity (v_p). Both are functions of wavelength. Phase velocity is:

$$v_p(\lambda) = \frac{c}{n_p(\lambda)} \tag{11.1}$$

where n_p is the normal tabulated refractive index at wavelength λ and c is the speed of light. The group velocity and index are given by:

$$v_g(\lambda) = \frac{c}{n_g(\lambda)} \tag{11.2}$$

$$n_g(\lambda) = n_p(\lambda) - \lambda dn_p(\lambda)/d\lambda \tag{11.3}$$

For silica-based fibers in the infrared $n_g \approx n_p$ so we can simply assume that the wave in the fiber is slowed by the phase index, giving the propagation velocity as:

$$\frac{2.9979 \times 10^8}{1.47} \approx 2 \times 10^8 \, m/s \tag{11.4}$$

and a "slowness" of about 5 ns/m. The different lengths of monitor-arm and sensor-arm fibers give us the possibility to adjust the two reflections to be in quadrature at the detector. If this is done, then a synchronous detector aligned with the signal will show zero, or at least very greatly suppressed response to the disturbing monitor-arm reflection.

For quadrature of the two waveforms, they need a relative phase displacement of $\pi/2$, or one quarter of a modulation period. Equating the time delay that we need, $\delta t = 1/(4 \cdot f_{mod})$ to the group delay in the extra fiber of length L_s, we obtain:

$$L_s = \frac{c}{8 n_g f_{mod}} \tag{11.5}$$

With a 100 kHz modulation frequency this corresponds to a fiber of length $L_s = 255$ m. With the low loss of fiber (they are the highest time-bandwidth product delay lines available), it is often convenient to pad out the length of one line to achieve this quadrature characteristic. Suppression of the monitor reflection response by a further 100 : 1 should be straightforward with this technique. In practice it is also possible to vary the modulation frequency under computer control. This would be very useful to accurately and actively null out the monitor-arm response, allowing for length changes in the fiber due to substitution of alternative connectorized fiber lengths. The combination of frequency synthesizer and lock-in amplifier is a very powerful workstation for fiber sensor development and use. Of course, with a two-channel measurement system like this, you will be asking "Why not measure the power of the monitor reflection in the same way, and do without the monitor photodiode entirely?" This is certainly an attractive approach. We have then reduced the system to a single photoreceiver, although we have a true signal plus reference measurement. The two-phase lock-in amplifier can do it all.

There are many variations on this theme which have been published and even used in real applications. We could, for instance, use a partial reflector half-way along the sensor fiber as the power monitoring reflection. This might reduce errors due to split ratio in the fiber coupler.

11.5 Spectral Frequency Analysis

Frequency coding is clearly capable of multichannel operation, but where more than two or three channels are required, the fabrication and adjustment of one synchronous detector per channel becomes inconvenient, even in the lab. We could of course use a single synchronous detector with a frequency synthesizer or just the individual source modulation generators connected to the reference input. In this way we could sequentially select the different frequency components. Alternatively, we might move from dedicated fixed-frequency synchronous detectors to full spectral analysis. After digitizing and storing a series of time-samples, the magnitudes of each frequency-coded channel can be calculated in software. Figure 11.7 shows a system in which the separation of different frequency signals is performed digitally, for example using an FFT (Fast Fourier Transform) process. The FFT algorithm has become so ubiquitous that it is easy to think that it is the only spectral analysis technique to use. For a few, known spectral components, however, the FFT is overkill, and a (slow) Fourier transform performed in the classic way may be faster. Just as in the case of lock-in detection, we can operate either with or without a phase refer-

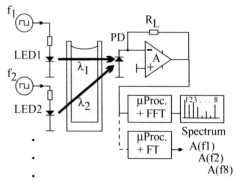

Figure 11.7 Extension to multiple channels is straightforward, although electronic frequency separation may require (slow) or fast Fourier transform analysis.

ence clock. If no references are available, then both sine and cosine transforms will have to be computed and added as sums-of-squares, in order to resolve the phase ambiguity. This is then essentially a digital or software implementation of the two-phase lock-in amplifier. Where large numbers of channels are to be monitored, this flexibility of digital spectral analysis is very enticing.

11.6 Sequence Analysis: Orthogonal Binary Coding

The idea of labelling a number of optical signals so that they can be sent through a single channel and separated at the end is closely related to the problem of coding and transmission for communications uses. Why can't we use all the available techniques from that field for instrument use? What is wrong, for example, with serial coding such as the RS232/V24 standard, where we can make use of all the UARTs and other cheap digital chips for coding and decoding? We could have the red-wavelength channel transmit a string of "A" characters, the blue channel a string of "B"s. Let's look at this actual example. The character "A" in ASCII is represented by 65_{dec} (01000001_{bin}) "B" by 66_{dec} (01000010_{bin}). These could be transmitted simultaneously, and we could look for just those sequences at the output. We need to multiply the aligned received signal by the binary number we expect, and add up the result. The problem is that "A" and "B" are not orthogonal, so an "A" will give some output from the "B" channel and vice versa. We can see this by translating 0s to −1, and multiply the "A" and "B" binary numbers:

$$\sum [A(-1\,1\,-1\,-1\,-1\,-1\,-1\,1) \times B(-1\,1\,-1\,-1\,-1\,-1\,1\,-1)] = 4.$$

On the other hand, if we had chosen the characters "U" (85_{dec}, 01010101_{bin}) and "i" (105_{dec}, 01101001_{bin}), we would have obtained a zero sum. These two characters are, in this representation, orthogonal. It is clear that not all characters are equal. If we choose the characters carefully, the electronics of serial transmission systems could in principle be used for instrumentation. This leads us back into the analysis of Walsh functions introduced in Chap. 5.

The sine/cosine series has been the basis of coding theory and signalling for the past century. This is partly because these functions form a complete, orthogonal set. By complete, we mean that with increasing number of terms, the mean-square error in approximation of a function approaches zero. By orthogonal, we mean that the integral of products of pairs of terms vanishes. However, sine waves are in many situations more difficult (i.e., more expensive) to generate than binary digital waveforms. Hence there is great interest in sets of binary waveforms which are also orthogonal. The Walsh functions form one set, which we have already used to synthesize a sine wave. These are binary functions which take on values of ±1, and are written WAL(n, N), where N takes the role of the time-base ordinate, and n the harmonic number. N is also the number of waveforms in the set, which should be a power of 2. We looked at a

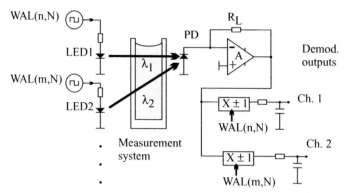

Figure 11.8 Walsh function transmission can be used to label many measurement channels.

few examples in Fig. 5.12 for $N = 16$. Some of these functions are the normal f, 2f, 4f etc. square waves, just as we have used for LED modulation up to now. Others, however, exhibit a nonuniform mark-to-space ratio. Corresponding to the *frequency* of a sine wave, Walsh functions are characterized by the number of transitions per timebase, or *sequency*. N.B. Mathcad's notation is slightly different. The function Walsh(n,m) returns the *m*th vector in the set of 2^n elements. For example, Walsh (3,5) returns $(1,-1,1,-1,-1,1,-1,1)$.

For multichannel transmission, each differently colored LED should be driven by a distinct Walsh function. At the receiver, in a manner equivalent to the sine-wave demodulation of Fig. 11.4 we multiply by the appropriate Walsh waveform and integrate (Fig. 11.8). The key point is that the multiplication is always by $+1$ or -1, which can be done with a simple switch. After integration, the detected magnitudes are obtained free from interchannel interference. A disadvantage of Walsh functions compared with the sines is that reference phasing is more difficult. With sine waves it is only necessary to shift one cycle at most to align the received waveform with the reference. With Walsh functions it may be necessary to shift one complete time-base. Nevertheless, they are worth investigating for multichannel use, as the electronic hardware is fast and efficient due to the use of binary functions. A further advantage of using the nonperiodic functions is that modulation energy is distributed over a wider range of frequencies. As we have seen, square-wave modulation puts energy into harmonics at f_{mod}, $3f_{mod}$, $5f_{mod}$ etc. With sine-wave-reference synchronous detection we determine the magnitude of the component at f_{mod}. If we are unlucky, and an interference source appears at this frequency, then severe intermodulation might occur, leading to large measurement errors. By contrast, with a Walsh coding system, source energy is spread over a wider range of frequencies, and synchronous detection detects simultaneously at all of them (as long as the receiver bandwidth is high enough). Now, strong interference at one specific frequency is expected to lead to lower error, as it only affects a small fraction of the channel bandwidth.

There are many ways to generate the Walsh functions. Where only eight channels are needed, it is simple to hard-code the 8-bit bytes into a micro-controller, and to read them out sequentially. The spectra of Fig. 5.4 were obtained in this way with sequential readout of memory locations in a Basic Stamp computer, which contains a PIC microprocessor. Henning Harmuth (1964) has given detailed theory of the numerical generation of Walsh functions, while Beslich (1973) has described solutions for Walsh generators using digital logic.

11.7   Time Multiplexing

11.7.1   Source-polling

One of the simplest multichannel techniques for a few or a few dozen channels, as long as a microcontroller is available in the measurement setup, is to time multiplex. The sources are illuminated in turn, and a common photo-receiver determines the intensity in that channel. If the multiplexing is performed thousands of times per second, this is equivalent to shifting the measurement to higher frequencies. The repetitive short-time samples give a periodic comb of passbands in the frequency domain. By spreading the information over a wide range, interference from specific, unfortunately-placed signals can be suppressed. Hence performance can be better than that of a narrow-band modulation/demodulation system.

11.7.2   Weighing designs

The problem with time division into a large number of time-slots is the poor use of the source's energy. As we have seen, the signal to noise ratio (S/N) is determined by the number of photons detected during a measurement, so arranging for each source to be off for most of the time is not the way to optimize performance. We need to have sources on for as much of the time as possible. Improvements can be made by illuminating *combinations* of sources together, and detecting the now larger composite signals. This is equivalent to the classic "weighing problem" of statistical analysis, and treated by Yates (1935). A short treatment is given in App. D. For now we limit ourselves to a brief description of the technique, and work through a tiny example. Assume that we want to measure the light intensities from three weak sources, arranged here as a vector,

$$\mathbf{I}_{\mathrm{true}} = \begin{bmatrix} 10.6 \\ 11.3 \\ 13.7 \end{bmatrix}$$

and the S/N is not good because of the receiver noise, which is normally dis-tributed. The detection variance, combined with the weak received intensities, gives a large relative error for each of the three measurements. Now let us perform a different set of three measurements, by first illuminating LEDs 1 and 3, then 2 and 3, then 1 and 2. The choice of on/off LEDs at each stage is given by subsequent rows of the so-called **S**-matrix:

$$\mathbf{S} = \begin{bmatrix} 1 & 0 & 1 \\ 0 & 1 & 1 \\ 1 & 1 & 0 \end{bmatrix}$$

We can conveniently use a microcontroller to sequentially illuminate the three LED groups. With each combination we digitize the integrated intensity as usual. The resulting measurements are:

$$\mathbf{M} = \begin{bmatrix} 24.3 \\ 25.0 \\ 21.9 \end{bmatrix}$$

Each combination provides one intensity value. Because the aggregate intensity of the three signals in each measurement is larger (about twice as large) than that of a single source, the relative error of the second set (**M**) will be reduced by comparison with set (I_{true}). To recover the individual channel intensities we multiply the vector **M** by the correct inverse **S**-matrix:

$$I_{\text{est}} = \mathbf{S}^{-1}\mathbf{M} = \begin{bmatrix} 0.5 & -0.5 & 0.5 \\ -0.5 & 0.5 & 0.5 \\ 0.5 & 0.5 & -0.5 \end{bmatrix} \begin{bmatrix} 24.3 \\ 25.0 \\ 21.9 \end{bmatrix} = \begin{bmatrix} 10.6 \\ 11.3 \\ 13.7 \end{bmatrix}$$

The improvement in S/N which is possible with only a few channels is rather modest. With $N = 3$ the gain of 1.22 is not impressive, but becomes interesting for larger N.

In particular, this technique has been put to very effective use in infrared spectroscopy. When dispersive spectroscopy is performed with narrow input and output slits, most of the source power is lost for most of the time. If instead the usual single slits are replaced by coding masks in the form of one-dimensional **S**-vectors, the improvement in light throughput can be large (about $N/2$ for one mask). Either input-only, output-only or both slits may be so replaced. Light passing through illuminates a single detector. To record the spectrum the mask is scanned just as in the recording of a direct spectrum. After recording, the true spectrum is recovered through a software matrix product. At a time when only single detectors were available, resolution and S/N of these so called Hadamard spectrometers were much superior to classic instruments. Harwit and Sloane (1979) provide a beautiful description of their design and operation.

With the development of large area detector arrays, which can detect photons in all spectral channels simultaneously for the full time of the measurement, one advantage of these Hadamard multiplexing techniques vanishes. A Hadamard mask at the input can still provide increased optical throughout from an extended source. The principles involved here are interesting and didactic for the whole of optical measurement. At the very least they emphasize the need to capture as much as possible of the source's available energy during a measurement period.

Calculations of Simple RC, LC Networks Using Complex Frequency Notation

A.1 Introduction

Resistor/capacitor and to a lesser degree resistor/capacitor/inductor networks form the basis of a huge variety of common laboratory electronic circuit techniques, including quiet and stabilized power supplies, noise filters, AC-coupling, differentiators, sample-and-holds and transmission lines. They also provide good models to calculate much of the behaviour of photodetection systems, from the simplest resistively-loaded photodiode to the most complex filtered transimpedance amplifier. We have seen that system optimization is a complex matter, with strong interactions between noise and bandwidth performance depending on the choice of the active elements and passive components used. This all suggests that we will not get far without convenient techniques to reliable calculate their performance over the frequency range of interest.

The choice of tools for this job is large, ranging from elaborate electronic simulation ("Spice") software, which gives the results without much feel for what is important, to rigorous analysis with pencil and paper, which is big on feel, but too tedious for all but the simplest circuits. Spreadsheets, general mathematical programming languages and symbolic software hold the middle ground. At different times it is useful to have all these tools available, with the convenience of one balanced against the fancy graphics of the other. Whichever package you use there is always something you wish it would handle better. What is probably much more telling than the choice of analysis tool, however, is one's fluency with it. It is often better to be an expert user of an outdated package than to stumble along in the latest version with all those arcane new features.

A.2 Complex Frequency Analysis

One consistent approach to formal calculation of photodetector circuits is the use of the complex frequency s. A few simple examples of its use will be treated

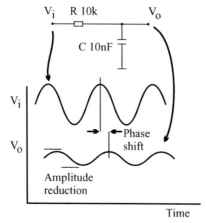

Figure A.1   The RC low-pass filter modifies both amplitude and phase as a function of frequency.

now. Using $j = \sqrt{-1}$ for the complex variable we define the complex (radian) frequency in terms of the real frequency f as:

$$s = j2\pi f \tag{A.1}$$

The impedances of a capacitor C and inductor L are then given in an obvious notation by:

$$Z_C = \frac{1}{sC} \tag{A.2}$$

$$Z_L = sL \tag{A.3}$$

The RC network of Fig. A.1 is a voltage divider, driven from a low-impedance source, whose output voltage V_o of course drops with increasing frequency, as the impedance of the capacitor decreases. We can immediately write the division ratio $A(s) = V_o/V_i = Zc/(R + Zc)$. This simplest low-pass filter can be described by a break or characteristic frequency f_c beyond which the output drops; f_c is defined as the frequency at which the impedances of capacitor and resistor are equal ($Zc = R$). However, note that even in this trivial circuit, some care is needed with such a statement.

In order to calculate the division ratio of the low-pass as a function of frequency we *could* go through the algebra:

$$A(s) = \frac{1/sC}{R + 1/sC} = \frac{1}{1 + sRC} \tag{A.4}$$

and then decompose $A(s)$ into magnitude $|A(f)|$ and phase $\varphi(f)$ terms:

$$A(s) = |A(f)|e^{j\varphi(f)} \tag{A.5}$$

$$|A(f)| = \frac{1}{\sqrt{1 + (2\pi f RC)^2}} \tag{A.6}$$

$$\varphi(f) = -\tan^{-1}(2\pi f RC) \tag{A.7}$$

However, this is pretty tedious algebra; even for a trivial case I find that if I have appropriate software, it is much better to avoid the symbolic evaluations of Eqs. (A.6) and (A.7) and work with Eq. (A.4). In a more complicated network the expression for the complex transfer function $A(s)$ may be fairly daunting, but carry the calculations through regardless. Only at the end, for example for plotting the graphs, do I convert to magnitude and phase using the built-in functions. Mostly I like to use a modern general purpose mathematical programming language for this, and in particular one which handles complex arithmetic in a straightforward and transparent way. Excel deals with complex numbers, but is a little cumbersome. Mathcad is a good example, which in addition is almost self-documenting. In Mathcad, obtaining the magnitude and phase of $A(s)$ involves no more than writing:

$20 \log(|A|)$ for the magnitude in dB referenced to unity voltage, and

$360/2\pi \arg(A)$ for the phase in degrees.

With the component values shown (10 k, 10 nF), $f_c = 1592\,\text{Hz}$, and the impedance and output voltage at $f = f_c$ can be calculated to be:

$$Z_C = 0 - j\,10\,k\Omega \tag{A.8}$$

$$V_o = \frac{1}{2} - j\,\frac{1}{2} \tag{A.9}$$

With these values for the two components of our circuit, one pure real and the other pure imaginary, it is hard to call Z_c and R really "equal." If we do concede that R and Z_c are in some sense both equal to $10\,k\Omega$, then we must not take this to imply that the output voltage is half the input voltage! The problem lies in the complex nature of the variables. Above we described the complex voltage and impedance in real and imaginary parts, but what does this mean in the laboratory?

A.3 TRY IT!

The best way to see these effects is to try it. If Figure A.1 is set up as shown, with the scope triggered from V_i on Channel A, the variation of V_o with frequency can be observed on Channel B. However, we don't see real and imaginary parts, just "the waveform." Instead of real and imaginary parts, we could equivalently represent the same signal by a (real) magnitude M and a (real) phase angle φ, which are visible on the scope display. For a $V_i = 1\,\text{V}$ (peak) input signal the output voltage magnitude and phase are:

$$M = \sqrt{[\text{Re}^2(V_o) + \text{Im}^2(V_o)]} \tag{A.10}$$

$$\varphi = -\tan^{-1}\left[\frac{\text{Im}(V_o)}{\text{Re}(V_o)}\right] \tag{A.11}$$

where $\text{Re}(V_o)$, $\text{Im}(V_o)$ are the real and imaginary parts of V_o. At low frequencies input and output signals are equal in magnitude and in phase. As the frequency is raised, V_o starts to reduce in magnitude, but also to shift to the right (i.e., to later times). What is seen on the scope is the magnitude M of the signal, shifted in phase by φ with respect to the input.

At the characteristic frequency $f_c \approx 1.592\,\text{kHz}$ we can calculate that the in-phase (Re) component has been reduced to one-half, however the magnitude seen on the scope is not one-half, but $\sqrt{\left(\frac{1}{2}^2 + \frac{1}{2}^2\right)} = 0.707$. At f_c it is not the *magnitude* that is one-half (the magnitude has been reduced to $1/\sqrt{2}$ of the input voltage), but the *power*. If the scope is triggered on V_i, we can also roughly see the relative phase, which at f_c is $-45°$.

Let's work this through for the simple low-pass RC network, as well as a high-pass RL network. These analyses are written in Mathcad, although this is only one possibility. It is convenient for this because, as the program is self-documenting and looks almost like maths notation, it can be understood by anyone familiar with the mathematics, even if not directly familiar with Mathcad. The magnitude and phase results, or what would be seen on the scope in the TRY IT! are shown in Fig. A.2 in the Mathcad analysis. At f_c the transfer gain is 0.707 or in dB: $20\log_{10}(0.707) = -3\,\text{dB}$.

Figures A.3a and A.3b show straight-line approximations to the continuous curves. The magnitude is drawn constant to f_c, and then decreases with a fixed slope of $-20\,\text{dB/decade}$ of frequency. The phase is zero from the lowest frequencies up to $f_c/10$, after which it decreases at $-45°/\text{decade}$ up to $10f_c$, after which it remains at $-90°$. The straight line approximations are easy to remember, and accurate enough for most design purposes. At the two break frequencies of $f_c/10$ and $10\,f_c$, the phase error is only 6°, negligible for most purposes.

The equivalent Mathcad calculation for an RL network ($100\,\Omega$, $100\,\mu\text{H}$) is shown in Fig. A.4.

The complex frequency notation is a convenient, consistent approach to these simple calculations, which can be extended without difficulty to much more complicated networks. In many cases circuit calculations can be done without the use of Spice software. This gives a better feel for the effects of component and topology changes, and is often a better way to optimize the electronic designs. The disadvantage is that you lose the drag-and-drop circuit construction feature.

CALCULATIONS OF SIMPLE RC NETWORKS
rcsa.mcd

Plotting parameters: $fmin := 1.0 \cdot 10^1$ $fmax := 1.0 \cdot 10^7$ $N := 96$ $n := 0..\,N - 1$

Define the complex frequencies: $f_n := 10^{\left[\frac{n \cdot (\log(fmax) - \log(fmin))}{N} + \log(fmin)\right]}$ $s_n := 2j \cdot \pi \cdot f_n$

Low-pass RC network: $R := 10 \cdot 10^3$ $C := 10 \cdot 10^{-9}$

Char. frequency: $fc := \dfrac{1}{2 \cdot \pi \cdot R \cdot C}$ $fc = 1.592 \cdot 10^3$

Impedance of capacitor and transfer function: $Zc_n := \dfrac{1}{C \cdot s_n}$ $A_n := \dfrac{Zc_n}{R + Zc_n}$

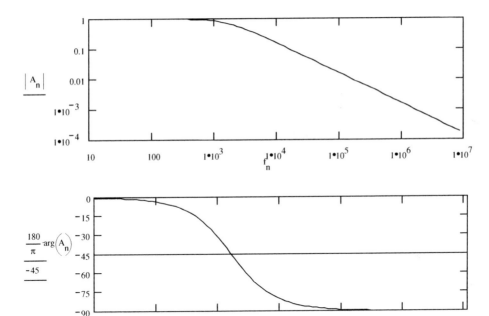

At the characteristic frequency: $f1 := fc$ $s1 := 2j \cdot \pi \cdot f1$

Impedance of capacitor: $Zc1 := \dfrac{1}{C \cdot s1}$ $Zc1 = -1 \cdot 10^4 \, j$

Transfer function: $Vo := \dfrac{Zc1}{R + Zc1}$ $Vo = 0.5 - 0.5j$

Figure A.2 Mathcad printout. Magnitude and phase response of the RC low-pass filter.

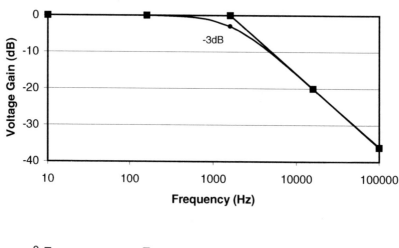

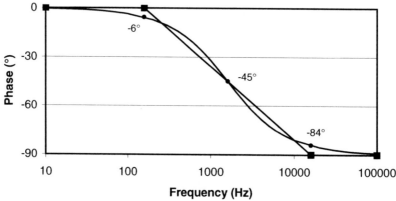

Figure A.3 Straight line approximations to the low-pass filter magnitude and phase result in little error, and are easy to remember.

A.4 Opamp Calculations

Calculations of the gain and phase of operational amplifier circuits are only slightly more complicated than the simple RC network treated above. As long as we are rigorous in application of the frequency dependent complex quantities, the answers just come out in the working. Here we will go through a few more examples.

A.5 Basic Opamp Circuits

Assuming an ideal opamp in the circuits of Fig. A.5, the two basic inverting and noninverting configurations exhibit the closed loop gains $A_{vc} = V_o/V_i$ given by $-Z_f/Z_1$ and $1 + Z_f/Z_1$ respectively. If the boxed components are just pure resistors, we obtain the equivalent well-known gain expressions $-R_f/R_1$ and $1 + R_f/R_1$.

CALCULATIONS OF SIMPLE RL NETWORKS
rcsb.mcd

Plotting parameters: $fmin := 1.0 \cdot 10^1$ $fmax := 1.0 \cdot 10^7$ $N := 96$ $n := 0 .. N - 1$

Define the complex frequencies: $f_n := 10^{\left[\dfrac{n \cdot (\log(fmax) - \log(\quad))}{N} + \log(fmin)\right]}$ $s_n := 2j \cdot \pi \cdot f_n$

High-pass RL network: $R := 1.0 \cdot 10^2$ $L := 100 \cdot 10^{-6}$

Char. frequency: $fc := \dfrac{R}{2 \cdot \pi \cdot L}$ $fc = 1.592 \cdot 10^5$

Impedance of inductor and transfer function: $ZL_n := L \cdot s_n$ $A_n := \dfrac{ZL_n}{R + ZL_n}$

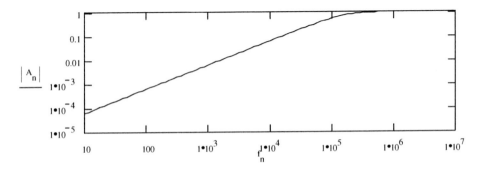

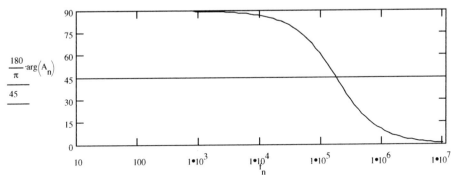

At the characteristic frequency: $f1 := fc$ $s1 := 2j \cdot \pi \cdot f1$

Impedance of inductor: $ZL := L \cdot s1$ $ZL = 100j$

Transfer function: $Vo := \dfrac{ZL}{R + ZL}$ $Vo = 0.5 + 0.5j$

Figure A.4 (Mathcad printout). Magnitude and phase response of the RL high-pass filter.

(a)

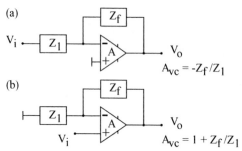

$$A_{vc} = -Z_f/Z_1$$

(b)

$$A_{vc} = 1 + Z_f/Z_1$$

Figure A.5 Opamp circuit gain and phase are easily determined even for arbitrary input and feedback impedance networks, as long as there is no connection to ground.

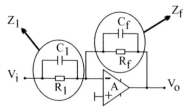

Figure A.6 Example inverting opamp circuit for analysis.

Here we just emphasise that the boxed components can be (almost) any network containing linear elements such as resistors, capacitors and inductors, arranged in arbitrary complexity. The gain expressions as written remain valid. They are also valid if the Z are complex.

For example, if in Fig. A.5a Z_1 is a parallel combination of capacitor C_1 and resistor R_1, and Z_f is also such a combination C_f, R_f, as in Fig. A.6, the gain A_{vc} is just:

$$A_{vc} = \frac{-Z_f}{Z_1} \tag{A.12}$$

where

$$Z_1 \equiv R_1 \| C_1 \quad \text{(parallel combination of } R_1, C_1\text{)}} \tag{A.13}$$

$$Z_f \equiv R_f \| C_f \quad \text{(parallel combination of } R_f, C_f\text{)}} \tag{A.14}$$

Let's work it through using the complex notation:

i) Define the complex frequency: $s = j2\pi f$ (A.15)

ii) Calculate the complex component impedances:

$$Z_{C1} = \frac{1}{sC_1} \tag{A.16}$$

$$Z_{Cf} = \frac{1}{sC_f} \tag{A.17}$$

iii) And the feedback element impedances:

$$Z_1 = \frac{R_1 Z_{C1}}{R_1 + Z_{C1}} \tag{A.18}$$

Plotting parameters: $\quad fmin := 1.0 \cdot 10^0 \qquad fmax := 1.0 \cdot 10^5 \qquad N := 100 \qquad n := 0 .. \ N - 1$

Define the complex frequencies: $\quad f_n := 10^{\left[\frac{n \cdot (\log(fmax) - \log(fmin))}{N} + \log(fmin)\right]} \qquad s_n := 2j \cdot \pi \cdot f_n$

Amplifier noise sources: $\quad vn := 10 \cdot 10^{-9} \qquad in := 1.1 \cdot 10^{-15}$

Photodiode: $\quad Cin := 16 \cdot 10^{-12} \qquad Zin_n := \dfrac{1}{s_n \cdot Cin} \qquad Id := 300 \cdot 10^{-12}$

Feedback arm: $\quad Cf := 1 \cdot 10^{-12} \qquad Rf := 100.0 \cdot 10^6 \qquad ZCf_n := \dfrac{1}{s_n \cdot Cf}$

Total Z of feedback arm: $\quad Zf_n := \dfrac{Rf \cdot ZCf_n}{Rf + ZCf_n} \qquad m := 60 \qquad f_m = 1 \cdot 10^3$

Feedback resistor internal current noise: $\quad IR_n := \dfrac{4 \cdot 10^{-12}}{\left(\dfrac{Rf}{10^3}\right)^{0.5}} \qquad\qquad IR_m = 1.3 \cdot 10^{-14}$

Feedback resistor output thermal noise voltage: $\quad VR_n := IR_n \cdot Zf_n \qquad\qquad \left| VR_m \right| = 1.1 \cdot 10^{-6}$

Amp current noise output voltage: $\quad VIN_n := in \cdot Zf_n \qquad\qquad \left| VIN_m \right| = 9.3 \cdot 10^{-8}$

Dark current shot noise output voltage: $\quad VID_n := 0.57 \sqrt{Id \cdot 10^6 \cdot 10^{-12}} \cdot Zf_n \qquad \left| VID_m \right| = 8.4 \cdot 10^{-7}$

Multiplied amp output voltage noise: $\quad VN_n := vn \cdot \left(\dfrac{Zf_n + Zin_n}{Zin_n}\right) \qquad \left| VN_m \right| = 9.1 \cdot 10^{-8}$

Total output signal: $\quad Ip := 8.0 \cdot 10^{-14} \qquad Sig_n := Ip \cdot Zf_n \qquad \left| Sig_m \right| = 6.8 \cdot 10^{-6}$

Signal shot noise output voltage: $\quad VIP_n := 0.57 \sqrt{Ip \cdot 10^6 \cdot 10^{-12}} \cdot Zf_n \qquad \left| VIP_m \right| = 1.4 \cdot 10^{-8}$

Total output noise: $\quad Vtot_n := \left[\left(\left| VN_n \right|\right)^2 + \left(\left| VIN_n \right|\right)^2 + \left(\left| VID_n \right|\right)^2 + \left(\left| VIP_n \right|\right)^2 + \left(\left| VR_n \right|\right)^2\right]^{0.5}$

$$\left| Zf_m \right| = 8.5 \cdot 10^7$$

Output signal to noise: $\quad SN_n := \dfrac{Sig_n}{Vtot_n}$

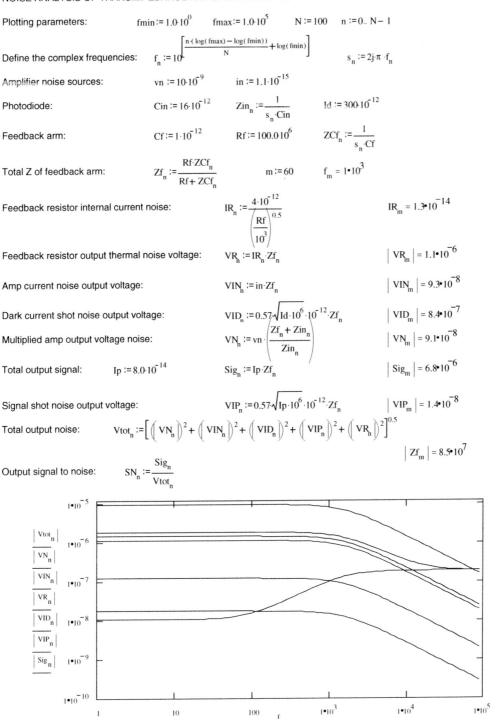

Figure A.7 (Mathcad printout). Noise analysis of the transimpedance amplifier.

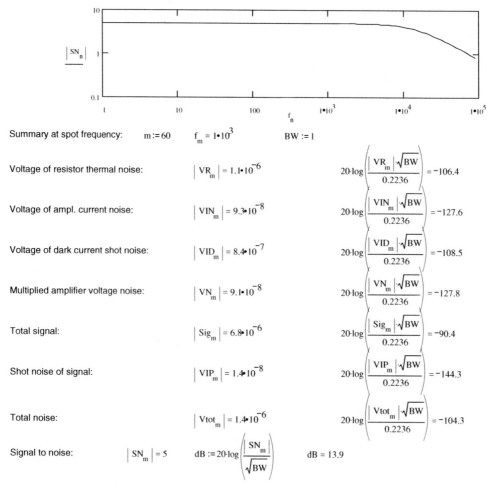

Summary at spot frequency: $m := 60$ $f_m = 1 \cdot 10^3$ $BW := 1$

Voltage of resistor thermal noise: $\left| VR_m \right| = 1.1 \cdot 10^{-6}$ $20 \cdot \log\left(\dfrac{\left| VR_m \right| \cdot \sqrt{BW}}{0.2236} \right) = -106.4$

Voltage of ampl. current noise: $\left| VIN_m \right| = 9.3 \cdot 10^{-8}$ $20 \cdot \log\left(\dfrac{\left| VIN_m \right| \cdot \sqrt{BW}}{0.2236} \right) = -127.6$

Voltage of dark current shot noise: $\left| VID_m \right| = 8.4 \cdot 10^{-7}$ $20 \cdot \log\left(\dfrac{\left| VID_m \right| \cdot \sqrt{BW}}{0.2236} \right) = -108.5$

Multiplied amplifier voltage noise: $\left| VN_m \right| = 9.1 \cdot 10^{-8}$ $20 \cdot \log\left(\dfrac{\left| VN_m \right| \cdot \sqrt{BW}}{0.2236} \right) = -127.8$

Total signal: $\left| Sig_m \right| = 6.8 \cdot 10^{-6}$ $20 \cdot \log\left(\dfrac{\left| Sig_m \right| \cdot \sqrt{BW}}{0.2236} \right) = -90.4$

Shot noise of signal: $\left| VIP_m \right| = 1.4 \cdot 10^{-8}$ $20 \cdot \log\left(\dfrac{\left| VIP_m \right| \cdot \sqrt{BW}}{0.2236} \right) = -144.3$

Total noise: $\left| Vtot_m \right| = 1.4 \cdot 10^{-6}$ $20 \cdot \log\left(\dfrac{\left| Vtot_m \right| \cdot \sqrt{BW}}{0.2236} \right) = -104.3$

Signal to noise: $\left| SN_m \right| = 5$ $dB := 20 \cdot \log\left(\dfrac{\left| SN_m \right|}{\sqrt{BW}} \right)$ $dB = 13.9$

Figure A1.7 *Continued.*

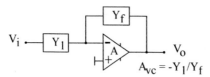

Figure A.8 Alternative representation in terms of input and feedback admittances.

$$Z_f = \frac{R_f Z_{Cf}}{R_f + Z_{Cf}} \tag{A.19}$$

iv) And the gain becomes $A_{vc}(f) = \dfrac{-Z_f}{Z_1} = \dfrac{R_f Z_{Cf}/(R_f + Z_{Cf})}{R_1 Z_{C1}/(R_1 + Z_{C1})}$ $\hspace{1cm}$ (A.20)

and substituting for Z_{c1} and Z_{cf} this simplifies to:

$$\frac{R_f(1 + sR_1C_1)}{R_1(1 + sR_fC_f)} \tag{A.21}$$

v) Plot gain and phase:

$\hspace{1cm}$ $20\log(|A_{vc}|)$ $\hspace{1.5cm}$ for the magnitude in dB (referred to $1\,V$ input)

$\hspace{1cm}$ $360/2\pi \arg(A_{vc})$ $\hspace{0.5cm}$ for the phase in degrees.

An alternative to this bottom-up approach is to use the symbolic substitution tools of some software packages to move from the top-down. In this approach we start by writing the opamp circuit gain expression: $A_{vc} = -Z_f/Z_1$, and then substitute for Z_f and Z_1 from Eq. (A.18), Eq. (A.19). Then we substitute the expressions for Z_{C1} and Z_{Cf} and so on. When all the substitutions have been made, the resulting expression can usually be simplified using the symbolic calculation engine. As noted, performing the symbolic simplification of Eq. (A.21) is not really necessary. We can instead just build up the large expressions and then plug in the values and calculate gain and phase at the end. This is the technique I find simplest. Following is an example of a noise calculation done in this way.

A.6 Transimpedance Noise Calculation

In Chap. 3 we looked at the calculation of noise and its frequency dependence in a transimpedance amplifier. These are not difficult calculations, but tedious for anything but a couple of values. We can use the software to help, and most importantly to show which contributions are dominant. Here is an example in Mathcad. Figure A.7 shows how the various terms vary with frequency.

A.7 Admittance Treatment

An alternative treatment of the feedback gain expression of Fig. A.5a is shown in Fig. A.8. Here the same components as before with impedances Z have just been relabelled with their admittances Y. The admittance is the reciprocal of the impedance. Just as the impedance Z can be split into its real and imaginary parts:

$$Z = R + jX \tag{A.22}$$

so can the admittance $Y = 1/Z$ be expressed as:

$$Y = G + jB \tag{A.23}$$

Now the gain of the inverting amplifier can be expressed, just as simply as with impedances, as:

$$A_c = -\frac{Y_{fl}}{Y_{tf}} \tag{A.24}$$

Note that the feedback elements seem to be the "wrong way up," but this is actually correct, as can easily be seen when the elements are pure resistors. It might appear that this is an unnecessary complication to the analysis, when impedances have up to now worked perfectly well. However, we will see presently that they are extremely useful.

The first area where an admittance treatment can be useful is in the analysis of complicated networks, but as in the case of frequency-dependent complex impedances, it is important here too to use a convenient language such as Mathcad for handling the complex admittances. It works as follows. Simply forming the reciprocal of a (complex) impedance Z correctly calculates the complex admittance Y. Further, we can switch between the two domains with the reciprocal operator, changing from ohms to mhos (now: S for Siemens) to have the correct units. Secondly, we know that for two components Z_1, Z_2 in series the combination impedance is trivially $Z_1 + Z_2$ even if the impedances are complex. If they are in parallel we must calculate $Z_1Z_2/(Z_1 + Z_2)$ which can lead to very messy expressions. By contrast, in the admittance world *parallel* admittances simply add together, while series connections require us to calculate $Y_1Y_2/(Y_1 + Y_2)$. By flipping back and forth between impedance and admittance domains, we can have the best of both worlds, and often get by just *adding* variables. This is even possible with hand calculation using a complex-arithmetic calculator. It is often possible to make a much "cleaner" development of the mathematics of large electrical networks by repeatedly flipping domains.

A.8   Grounded Feedback Networks

Earlier we stated that for the simple ratio-of-impedances expressions to be valid, the components in Fig. A.5 could be almost any impedance network; now we must clarify this caveat. For the simple feedback expressions to be valid, the feedback component is not allowed to have a connection to ground. Where a ground connection does exist, which is in practice often very useful, a unique impedance cannot be defined, and the simple ratio expressions cannot be used. The solution is to use instead admittances, or more precisely the short

(a)

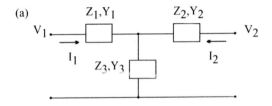

(b)

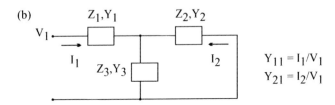

$$Y_{11} = I_1/V_1$$
$$Y_{21} = I_2/V_1$$

(c)

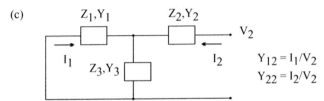

$$Y_{12} = I_1/V_2$$
$$Y_{22} = I_2/V_2$$

Figure A.9   Definition of the forward and reverse transfer admittances needed in opamp gain calculations.

circuit transfer admittances. With these, the expression for the gain A_{vc} will be correct.

Consider the three-terminal network of Fig. A.9a, in which the boxes represent arbitrary networks of elements with impedances Z_i or equivalent admittances $Y_i = 1/Z_i$. In general, these will be completely described by two (possibly frequency dependent) currents and two voltages. The *input* admittance Y_{11} is just the ratio of input current to input voltage (units of S). The *transfer* admittance Y_{21} is the ratio of output current to the input voltage. The expressions we need are those obtained with shorted networks. Looking into the network whose output is shorted to ground (b), $Y_{21} = I_2/V_1$ is the reverse transfer admittance. similar expressions can be written for a network whose input is shorted, as in (c). If in addition all elements used in the boxes are passive, that is, resistors, capacitors, diodes, transformers etc. then $Y_{11} = Y_{22}$ and $Y_{12} = Y_{21}$. This is a big simplification. The transfer admittance can then be calculated with either end shorted; one configuration may be simpler to analyze that the other. As long as we can write down the transfer admittances, these can be used to calculate the frequency-dependent gain of Fig. A.10 even if they contain grounded networks. We have $A_{vc} = -Y_{t1}/Y_{tf}$ where Y_{t1} and Y_{tf} are the transfer admittances of the input and feedback networks respectively.

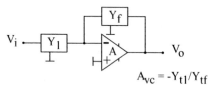

Figure **A.10** Inverting amplifier with grounded networks.

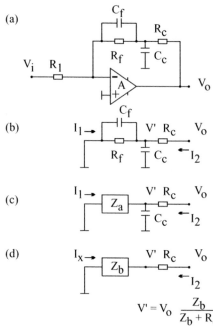

Figure **A.11** Step-by-step decimation of the feedback network admittances.

Let's take a simple but useful example. The transimpedance amplifier of Fig. A.11 has a feedback resistor R_f with a significant parasitic capacitance C_f. In series with this we have connected a low-pass network R_cC_c. On the input side a 1nA photocurrent is represented by a voltage of 1V in series with a 1GΩ impedance (10^{-9}S admittance).

In Fig. A.11 (b) we extract the feedback network and short the input side to ground. The task then is to calculate the transfer admittance, or the ratio of I_1 to V_o. It is usually possible by inspection to walk through the network from one end to the other. For example, in (b) to (d) in Fig. A.11:

i) Combine C_f and R_f into $Z_a = Z_{cf}/(Z_{cf} + R_f)$ where $Z_{cf} = 1/sC_f$

ii) Combine Z_a with C_c to give $Z_b = Z_{cc}Z_a/(Z_{cc} + Z_a)$ where $Z_{cc} = 1/sC_c$

iii) Calculate intermediate voltage $V' = V_oZ_b/(Z_b + R_c)$

iv) Calculate input current $I_1 = V'/Z_a$

v) Calculate transfer admittance $Y_{tf} = I_1/V_o$ (N.B. not I_x/V_o).

You can either work out the algebra to give a large explicit expression for Y_{tf}, or in a language like Mathcad just carry through the variables to give the answer. Let's try this latter approach.

We start with some specific values for a sensitive receiver, and calculate the frequency response of this photo-receiver:

$$R1 = 1\,G\Omega \quad R_f = 100\,M\Omega$$
$$C_f = 1\,pF \quad R_c = 100\,M\Omega \quad C_c = 1\,pF$$

The Mathcad analysis could look as follows.

As we learned in Chap. 2, the transimpedance and hence the output voltage will start to decrease above a characteristic frequency $f_c = 1/2\pi R_f C_f = 1592\,Hz$. The first part of the program calculates this, by setting the $R_c C_c$ time constant to a negligibly small value. The curves for phase and magnitude response are shown in Fig. A.12, lower curve. It illustrates the difficulty of obtaining adequate bandwidth from high transimpedance values when even tiny values of parasitic capacitance are present.

In the second section of the program we change the low-pass component values to: $R_c = 10\,k\Omega$ $C_c = 5\,nF$. The response changes to the upper curves in the Figure. The phase shift and the drop in sensitivity with frequency have been greatly reduced and the effective bandwidth extended. This compensation has been deliberately made imperfect, to better show the effect on gain. If R_c is changed again to $20\,k\Omega$, the sensitivity becomes independent of frequency, at least out to a new limit which would be given by the amplifier's open-loop gain. Perfect compensation is (theoretically) obtained when $R_f C_f = R_c C_c$.

This technique of compensation is very common, as it is used in every 10× oscilloscope probe (Fig. A.13). These are used to reduce the resistive loading on the circuit under test, but more importantly to reduce the capacitive loading of the 1 MΩ scope input. The 9 MΩ probe resistor is used as a resistive attenuator in conjunction with the 1 MΩ input resistance of the oscilloscope to give the 10× signal reduction. However, the scope input also has a parallel capacitance of typically 10 pF, which at 10 MHz, for example, would have an impedance of 1592 Ω, much lower than the 1 MΩ impedance expected. This would lead to a disastrously restricted bandwidth. This can be compensated with the addition of another capacitance across the probe resistor. With about 1.1 pF across the probe resistor, usually adjustable via a small screwdriver slot in the probe, such that the time-constant of the probe matches the time constant of the scope input ($R_1 C_1 = R_i C_i$), the probe will be compensated and give the best transient response. Just as adjustment of the probe compensation can give sluggish response to a transient voltage change, or a large overshoot, so the compensation for transimpedance capacitance must be adjusted for best transient response.

Although such a compensation can be very useful to increase detection bandwidth with high value transimpedances, it cannot be as perfect as the admittance analysis above suggests. This is because the model we have chosen for the physical transimpedance resistor, namely a simple parallel RC combination, is itself only an approximation. Certainly, adding the parasitic capacitance

ADMITTANCE ANALYSIS OF COMPENSATED TRANSIMPEDANCE AMP. admit1a.mcd

Plotting parameters: $fmin := 1.0 \cdot 10^1$ $fmax := 1.0 \cdot 10^6$ $N := 50$ $n := 0 .. N - 1$

Frequency arrays: $f_n := 10^{\left[\frac{n \cdot (\log(fmax) - \log(fmin))}{N} + \log(fmin)\right]}$ $s_n := 2j \cdot \pi \cdot f_n$

Uncompensated version: $Rf := 100 \cdot 10^6$ $Rc := 1.00$ $Vo := 1.0$

$Cf := 1 \cdot 10^{-12}$ $Cc := 1.0 \cdot 10^{-12}$

Calculate impedances: $ZCf_n := \dfrac{1}{s_n \cdot Cf}$ $ZCc_n := \dfrac{1}{s_n \cdot Cc}$

Merge Cf, Rf, Cf, Cc: $Za_n := \dfrac{ZCf_n \cdot Rf}{\left(ZCf_n + Rf\right)}$ $Zb_n := \dfrac{ZCc_n \cdot Za_n}{\left(ZCc_n + Za_n\right)}$

Intermediate voltage: $Vdash_n := \dfrac{Vo \cdot Zb_n}{Zb_n + Rc}$ Input current: $I_n := \dfrac{Vdash_n}{Za_n}$

Input & feedback admittances : $Yi_n := 10^{-9}$ $Yf_n := \dfrac{I_n}{Vo}$ Closed loop gain: $A_n := -\dfrac{Yi_n}{Yf_n}$

(Almost) compensated version: $Rf := 100 \cdot 10^6$ $Rc := 10000$ $Cf := 1 \cdot 10^{-12}$ $Cc := 5 \cdot 10^{-9}$ $Vo := 1.0$

Calculate impedances: $ZCf_n := \dfrac{1}{s_n \cdot Cf}$ $ZCc_n := \dfrac{1}{s_n \cdot Cc}$

Merge Cf, Rf, Za, Cc: $Za_n := \dfrac{ZCf_n \cdot Rf}{\left(ZCf_n + Rf\right)}$ $Zb_n := \dfrac{ZCc_n \cdot Za_n}{\left(ZCc_n + Za_n\right)}$

Intermediate voltage: $Vdash_n := \dfrac{Vo \cdot Zb_n}{Zb_n + Rc}$ Input current: $I_n := \dfrac{Vdash_n}{Za_n}$

Input & feedback admittances : $Yf_n := \dfrac{I_n}{Vo}$ $Yi_n := 10^{-9}$ Closed loop gain: $A2_n := -\dfrac{Yi_n}{Yf_n}$

Figure A.12   Gain and phase of the uncompensated and almost compensated transimpedance amplifier.

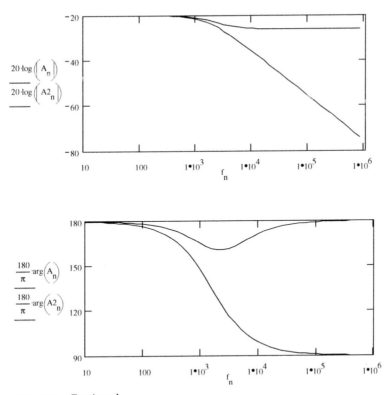

Figure A.12 *Continued.*

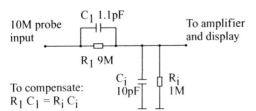

Figure A.13 The same compensation idea is usually applied in 10× scope probes.

provides a much better description of reality than a pure resistor alone, but we could do better still with the addition of further capacitive and inductive terms. In principle, these too could also be compensated, although the effort involved is probably more than that of obtaining an improved component or layout. If it is attempted, the admittance analysis will be a useful tool in simulating the results obtained.

We have shown here some of the tools needed to analyze the majority of circuits found in photo-receivers. Hopefully it has been shown that modern software which transparently handles complex arithmetic can be an enormous help in circuit calculations, even in more complicated situations such as transimpedance amplifiers with grounded feedback networks.

What Does the Spectrum Analyzer Measure

B.1 Introduction

The oscilloscope is great for observing more or less repetitive signals that are constant in time, but it gives only scant idea of the frequency content of any but the simplest of signals. The spectrum analyzer, on the other hand, gives little information on the time-variations, but a superbly detailed picture of the frequency content. Both are powerful and necessary tools for designing and debugging photo-measurement systems. The spectrum analyzer is used for noise and signal-to-noise measurements, for choosing modulation frequencies, for tracking down spurious signals and interferences. Indeed, there is a lot of merit in investigating all completed designs with a good spectrum analyzer, in order to see problems with spurious emissions and weak or out-of-band instabilities that go unnoticed on a time-domain display. This investigation must extend well beyond the notional signal frequency, as high-frequency instabilities can have a marked and difficult to understand effect on signal performance.

However, there is a lot of confusion about what the spectrum analyzer actually shows, which has been compounded by the new generation of digital analyzers, PC data-acquisition systems, and fast fourier transform (FFT) computing oscilloscopes. Hence it is useful to work though a few simple examples in pedantic detail (the simple ones we all get wrong!).

B.2 1V Peak Sine Wave

If a ± 1 V peak (2 V pk-pk) sine-wave voltage at 1 kHz (Fig. B.1) is applied to a conventional spectrum analyzer, what does the main spectral line read on the ordinate scale? To answer that we need to make a small detour into power definitions, and this means "dB." dBs describe ratios, so when using them we must always say what is the power to which the value is referenced. For historical reasons, in radio electronics the basis of the scale is usually defined as a root mean-square (rms) power of 1 mW dissipated in a resistance of $50\,\Omega$ (Take care.

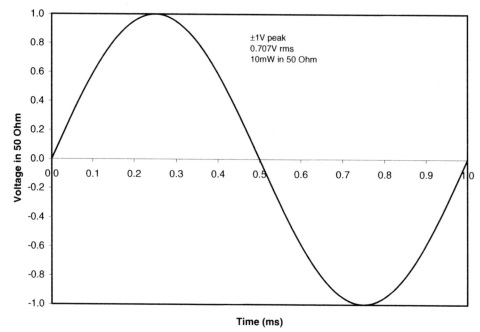

Figure B.1   One volt peak sine wave.

The audio engineers often use 600 Ω, but we will ignore them for now). This is defined as 0 dB referred to 1 mW, or 0 dBm. This is the basis of our scale of power ratios. One milliwatt in 50 Ω also implies a voltage of 0.2236 V rms, or 0.2236 × 2√2 = 0.6324 V pk-pk. Our ±1 V peak sine-wave voltage, with an rms voltage of 1/√2 = 0.7071 V, generates a power of 0.7071²/50 = 10 mW rms. In dBm this is calculated from the power ratio:

$$10 \log_{10}(10\,\text{mW}/1\,\text{mW}) = 10\,\text{dBm}$$

As the vertical axis of the spectrum analyzer is usually calibrated in dBm, this is the magnitude that should appear at the peak of the 1 kHz line. The displayed power in dBm can alternatively be calculated directly from the *voltage* as: $20 \log_{10}(0.7071\,\text{V}/0.2236\,\text{V}) = 10\,\text{dBm}$ Note the term "20," as we have now calculated a ratio of voltages, not powers.

It is a pity for much of modern laboratory electronics that the measurements are so usually phrased in terms of powers (which are rarely measured directly), in rms voltages (which are much harder to read on an oscilloscope than peak-to-peak voltages) and in matched 50 Ω source/detector impedances (when the majority of measurement are made at the low-impedance outputs of opamps and voltage references detected in a high impedance scope probe). Nevertheless, with a little care the pitfalls can be avoided.

TABLE B.1   **Synthesis of a 1 V peak Square-wave**

Harmonic (N)	Peak Voltage (V)	rms Voltage (V)	Power in 50 Ω (mW)	dBm
1	1.273	0.900	16.21	+12.1
3	0.424	0.300	1.80	+2.6
5	0.255	0.180	0.65	−1.9
7	0.182	0.129	0.33	−4.8
9	0.141	0.100	0.20	−7.0
11	0.116	0.082	0.13	−8.7
			19.3	

B.3   1 V Peak Square wave

As a just slightly more complex example, connect a 1 V peak 1 kHz square wave
to the spectrum analyzer. (Just check first that it can handle the power dissipation). The 1 kHz fundamental should read about +12 dBm. How do we know
that this is right? Well, it is straightforward to perform a Fourier analysis of
the square wave to determine the coefficients, that is, the amplitudes of each
of the harmonics that make up the square wave. This can be done analytically
and the result approximates our original square wave as: $4/\pi$ [sin x + 1/3 sin 3x
+ 1/5 sin 5x +...] These coefficients $4/\pi$, $4/3\pi$, $4/5\pi$ etc. of each harmonic are
the *peak voltages* of the sine waves that make up the symmetrical square wave.
I don't think Fourier mentioned rms values, which are of course $\sqrt{2}$ times lower,
and must be used to calculate the power in dBm delivered by each harmonic.
This is done in Table B.1.

Note that the sum of the powers delivered by the first six component sine
waves is 19.3 mW, close to the 20 mW we expect for a 1 V peak square wave or
a continuous 1 V in 50 Ω. Figure B.2 shows the first three sine waves with the
peak values from column two, their sum, and the original square wave that
made them. There is generally no true zero-frequency component on the spectrum analyzer's display. Hence we can offset the square wave on top of a DC
voltage without changing the display, as long as it is not overloaded.

B.4   Spectrum Calculation

With the wide availability of digital oscilloscopes, RS232, USB and wireless connection to your PC, and drag-and-drop waveform display and analysis, it is very
attractive to avoid paying for an expensive spectrum analyzer. Why not just
store a long time trace and calculate the spectra in software? It certainly is a
useful approach, but just as in understanding the units of a spectrum analyzer's
display, similar care must be applied when we use the discrete Fourier transform (DFT) to calculate spectra. One problem is that there are many different
definitions of these DFTs, usually calculated using the fast Fourier transform

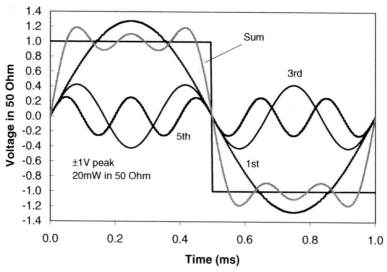

Figure B.2 One volt peak square wave decomposed into the first three harmonic components.

WHAT DOES THE FFT MEASURE?

Number of samples: $N := 128$ $n := 0 .. N - 1$ $ORIGIN \equiv 0$

Plot the time function: $V_n := \sin\left(\dfrac{2 \cdot \pi \cdot n}{N}\right)$ $nn := 0 .. \dfrac{N}{2} - 1$

Perform the Fourier transform (calculates the <u>peak</u> voltage): $U := \dfrac{2}{\sqrt{N}} \cdot fft(V)$

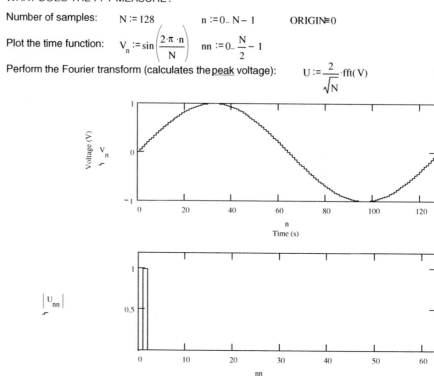

Figure B.3 Mathcad printout. Sampled sine wave and discrete FFT result.

Now a square wave: $Nm := 64$ $V_n := 0$ $s1 := 0.. Nm - 1$ $V_{s1} := 2$

Plot the time function: $V_n := V_n - 1$

Perform the Fourier transform,
and display the frequency domain: $U := \dfrac{2}{\sqrt{N}} \cdot fft(V)$

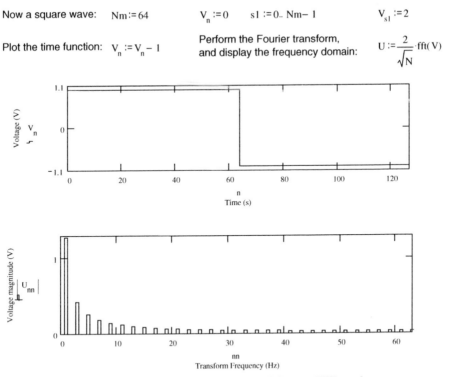

Figure B.4 Mathcad printout. Sampled square wave and discrete FFT result.

(FFT). Most of this is down to scaling constants for the forward and inverse transforms, but there are also different algorithms provided to save computational effort when you know something about the original dataset: is it real-only, is it symmetric about zero-time and so forth? The beautiful mathematical environment Mathcad, for example, provides at least four different algorithms. Each must be scaled correctly to obtain a meaningful measure of the component amplitudes of a transformed waveform. We should calculate the transformed vector **U** as follows, either as:

$$\mathbf{U} := \left(\frac{2}{\sqrt{N}} \right) \, fft(V) \quad \text{(lower case fft)}$$

or as:

$$\mathbf{U} := 2 \, FFT(V) \qquad \text{(upper case FFT)}$$

Here N is the number of points in the transform. The coefficients obtained in **U** represent the peak voltages of the component sine waves, corresponding to the peak voltages of Table B.1. For the case of a $\pm 1\,V$ sine wave and $\pm 1\,V$ square wave we could calculate as in Figs. B.3 and B.4.

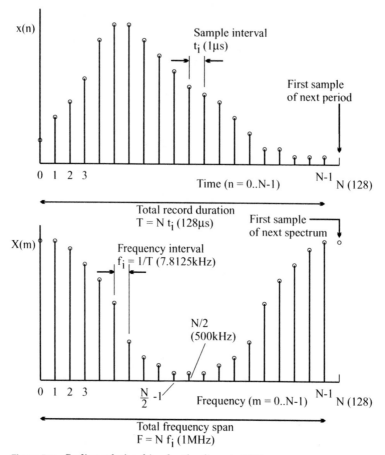

Figure B.5 Scaling relationships for the discrete FFT.

It will be seen that the amplitudes of the component sine waves of the ±1 V square wave agree roughly with the analytic calculation. These issues of scaling are not errors, but just down to the definitions used in each algorithm's implementation of the forward and inverse discrete Fourier transforms. However, incorrect scaling will lead to misleading quantitative results, so some care is needed.

Take care also with the time and frequency units. On the above plots we have only shown the data array ordinates, without any real unit scaling. If the input array is a time waveform of duration T seconds, then the units of the transformed frequency domain are really "cycles-per-T." Hence for our single-cycle of sine-wave time-waveform the transformed frequency is 1/T. If each time sample had a duration of $1\,\mu s$, the time-record would be $128\,\mu s$ long, and the first transformed frequency (after DC) would be 1 cycle-per-$128\,\mu s$, or

7.8125 kHz. The second frequency would be 15.625 kHz and so on, up to the maximum transformed frequency of $63 \times 7.8125 = 492.1875$ kHz. The next point $(64 \times 7.8125$ kHz) represents the Nyquist frequency (500 kHz), or the highest frequency that could be reproduced by a $1\mu s$ sample interval system. This is why the 128 (real) data-points in the time-record only give 64 valid points in the frequency-record.

Bluff Your Way in Digital Signal Processing

C.1 Introduction

This book has been concerned primarily with the analog domain of continuous-time detectors, opamps and RC networks. We have used it to design receivers, calculate their sensitivity, noise, bandwidth, rise time and so on. In the majority of cases, system performance will be defined by the analog domain characteristics of the early electronic stages. Nevertheless, at some point you will probably want to digitize your signals, and perhaps log a large number of measurements for analysis and reporting. It is often more convenient for this to perform some signal processing tasks in the digital domain, even if only averaging to reduce noise. There are many excellent books available on real digital signal processing (DSP). Here we touch on the absolute minimum requirements to survive in an increasingly digital world.

C.2 Sampling and Aliasing

Whenever we digitize an analog value, whether using a networked high-speed analog-to-digital converter and data logger or by writing down the value shown on our digital or analog voltmeter, we are entering the domain of "sampled data systems." The central problem of sampling is contained in Nyquist's theorem, which states that a continuous function can be perfectly reproduced from a set of sampled values, as long as the process captures at least two samples of the highest frequency component in the signal. Alternatively, this says that if we don't sample fast enough, we will see signals that are not really there. This is the problem of aliasing.

For a simple, concrete example, Fig. C.1 shows as a solid curve a 20 Hz sine wave input signal. For faithful reproduction, this would need to be sampled at >40 Hz (25 ms/sample) for perfect recovery. However, the points shown are samples taken at only $f_s = 22$ Hz (45.45 ms/sample). Without knowledge of the true 20 Hz waveform and with only the samples to deal with, we might be drawn

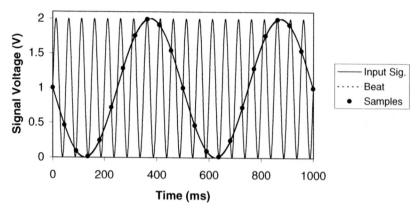

Figure C.1 Formation of an "alias" by under-sampling a 20 Hz sine wave. Sampling at 22 Hz the apparent beat occurs at 2 Hz.

to connect the sampled points with the line as shown and reconstruct a nice 2 Hz sine wave, but this is incorrect. There is no such component in the input signal! The 2 Hz apparent sine wave is an alias, which can be understood as the beat between the actual 20 Hz signal and our inadequate 22 Hz sampling. The sampling process mixes real components, by folding them about the Nyquist frequency $f_s/2$. The 20 Hz input is folded about the 11 Hz Nyquist frequency to give a 2 Hz alias, as plotted in Fig. C.2. If the Nyquist frequency is raised by sampling at 44 Hz there is no folding of signal frequencies into the 0–22 Hz band.

To avoid these artifacts, we must use an analog filter to remove all signal energy at or above the Nyquist frequency, at least to below the least significant bit (LSB) of the input analog-to-digital converter (ADC). This is not easy to do, and is often ignored in data acquisition systems of the PC variety. It is all too easy to program a 16-bit ADC in your favorite drag-and-drop programming environment, set up the sampling rate to ten samples per second, and wonder why there is an odd signal variation with a period of a few seconds in the logged data. If there is any 100/120 Hz light getting into your detector, it may be being aliased into the measurement channel as a slow beat.

To filter out the high frequency energy, you can add an RC low-pass filter before the input to your ADC, but with a cutoff slope of only 20 dB/decade, and perhaps 60 or 80 dB of suppression required, this may restrict the fastest signal change that can be followed to a very low frequency. For example, suppose we can manage a 10 Hz sample rate ($f_{Nyquist} = 5$ Hz) and need 60 dB suppression of aliased signal components (Fig. C.3). With a single RC filter 60 dB suppression is only obtained by setting the filter break-frequency to a point *three decades* below 5 Hz, or 0.005 Hz. However, this will give a system time response to a step input of about 0.35/0.005 = 70 s, giving us only one valid measurement every minute or so! What happened to our 10 Hz sample rate? The need to supress

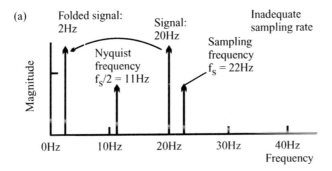

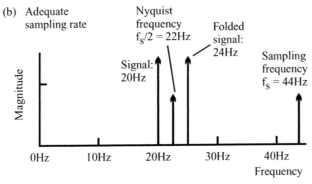

Figure C.2 (a) Frequency-domain depiction of under-sampling. The Nyquist frequency must be greater than the signal frequencies to avoid aliasing. (b) Adequate sampling rate allows correct reproduction of the 20 Hz signal.

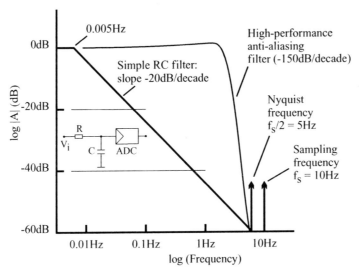

Figure C.3 Reduction of signal components at the Nyquist frequency to low levels requires setting the RC-filter break frequency to a very low value, leading to a low data rate. Anti aliasing filters push the break frequency close to the limit by using sharp cutoff slopes.

aliases has reduced our information-rate by about 1000 times. The conclusion is that a single RC filter is not very useful, and we should be looking for something with a faster cutoff. These are complex, so good antialiasing filters are fancy and expensive. They are usually designed to lose not three decades of signal frequency between filter cut-off and Nyquist frequency, but a fraction of a decade, by employing filters with very sharp cutoffs. In this way the effective signal bandwidth will be increased almost to the Nyquist value. For example, National Instruments' SCXI-1141 range of programmable low-pass filters use eighth order designs (8 RC time-constants) and achieve 135 dB/octave cutoff slopes and 80 dB attenuation. The filters are available with elliptical, Bessel or Butterworth responses, for fastest cutoff, best phase response, or flattest amplitude response in the passband respectively. Similar performance is available also in integrated circuit form. The Linear Technology LTC1546 is a programmable eighth order low-pass filter for 10 kHz to 150 kHz, with 100 dB attenuation at $2.5 \times f_{cutoff}$. This is impressive performance.

I have seen lots of elaborate data acquisition systems and dedicated dataloggers and even digital feedback control systems, but I have rarely come across a high performance antialiasing filter in an optics laboratory! In practice you are much more likely to find researchers either oblivious of the dangers, or thinking that aliasing only happens to DSP engineers. What can you do? First, do what you can to curtail the frequency response of your detection system. Have a look on the spectrum analyzer to see just how much power there is two or three decades above your signal variations of interest. If necessary, apply a compromize analog filter, perhaps a two- or three-pole filter. This can be especially necessary with transimpedance receivers, as noise peaking can give a noise power which is flat, or even increases with increasing frequency. Filtration here will at least reduce the energy at high frequencies. The next alternative is to increase the sampling rate. If the data-acquisition system is up to it, another ten-fold increase here can make analog filtration much easier. Of course you will also collect a lot more data that you don't really want, but this can be decimated using digital averaging routines as described below.

In summary, aliasing is not just the torment of DSP engineers. Even if the sampling-rate and filtration requirements are too daunting, at least an awareness of the effects can be very useful. Once you know that aliasing can turn high frequency interference into weird low-frequency signal variations, it is much easier to track down.

C.3   Data Smoothing in the Digital Domain

Many optical systems naturally provide measurement data at a faster rate than needed. For instance, a sensitive turbidimeter for drinking water applications may have a bandwidth of tens of hertz, although the slow variations in quality of the water supply require a reading only every ten minutes, requiring a bandwidth of only ≈ 0.0005 Hz. This low bandwidth could be set up, for example, in the analog demodulation filter of a synchronous detector. However, setting up and characterizing electronics with a millihertz bandwidth are inconvenient because of the sluggishness of the response. You may also want to switch the instrument later to an application with a far higher necessary bandwidth, such as in flow-mixing studies with a water supply spiked with small quantities of kaolin clay. It is therefore often convenient to leave the instrument fixed at a high bandwidth, and reduce the bandwidth in software in a post-processing step.

The simplest technique is the unweighted smoothing of the *sliding-filter*. This involves averaging a sequence of measurements, from the current value back in time. If the raw measured values are denoted V_i and the smoothed values U_i, we may, for example, calculate:

$$U_i = \frac{V_{i-4} + V_{i-3} + V_{i-2} + V_{i-1} + V_i}{5} \tag{C.1}$$

to average the five most recent values. As the next sample arrives, you recalculate the average with the shifted block. Alternatively, keep track of all the historical samples, subtract $V_{oldest}/5$ and add $V_{newest}/5$ at each step. In Mathcad this might be done as follows:

Number of points, counter: $N := 50$ $n := 0..N$

Set up input vector: $n1 := 0..5$ $n2 := 6..25$ $n3 := 26..50$

$V_{n1} := 0$ $V_{n2} := 1$ $V_{n3} := 0$ $U := V$

Calculate smoothing: $m := 4..N-1$

$U_m := (V_{m-4} + V_{m-3} + V_{m-2} + V_{m-1} + V)/5$

Figure C.4 shows the results. The square function is the input signal, the trapezoidal function the smoothed output. This is just as trivial to program using any other mathematical software, or even a spreadsheet. If the input data are arranged in column A, set up a cell in column B to be the average of the N = 5

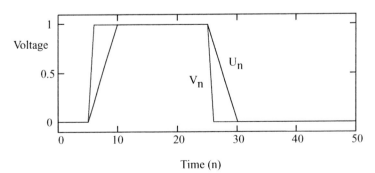

Time (n)

Figure C.4   Moving average digital filtration.

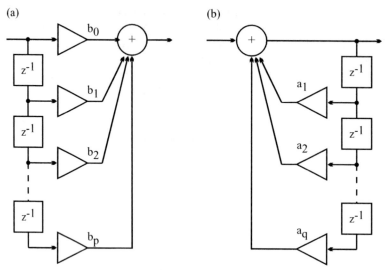

Figure C.5 Basic configurations of nonrecursive (a) and recursive (b) digital filters. z^{-1} represents a delay of one sample period.

cells above it and to the left, and copy the cell down the whole of column B. Remember to start the calculation from the Nth data point onwards. If N points are included in the sliding average, the first N-1 measurements do not have corresponding smoothed values.

The sliding averager is only one example of a discrete time filter. These are available in two main types, nonrecursive and recursive, represented in so-called Direct Form I in Fig. C.5. Each block marked z^{-1} represents a time delay of one sampling interval, while the a_i, b_i are the gain coefficients of each delayed sample. The algebra of the z-domain can be used to calculate the discrete time and frequency response of these filters. The transfer functions of the two filter types (a) and (b) are given respectively by:

$$H_1(z) = b_0 + b_1 z^{-1} + b_2 z^{-2} + \ldots + b_p z^{-p} \tag{C.2}$$

$$H_2(z) = \frac{1}{1 - (a_1 z^{-1} + a_2 z^{-2} + \ldots + a_q z^{-q})} \tag{C.3}$$

For the 5-point sliding averager we have:

$$H_1(z) = 1 + 1z^{-1} + 1z^{-2} + 1z^{-3} + 1z^{-4} \tag{C.4}$$

The sliding averager is simple, effective, and will reduce the standard deviation of a noisy dataset. However, it can lead to data artifacts, depending on the frequency content of the input variations. We can calculate the equivalent frequency response of this low-pass filter by Fourier transforming the time-domain filter, or equivalently by substituting $z = e^{2\pi j f}$ into Eq. (C.4), where f is the real frequency. Mathcad does this conveniently using its Symbolics|Variable| Substitute tool.

Z-domain representation: $H(z) := 1z^{-4} + 1z^{-3} + 1z^{-2} + 1z^{-1} + 1z^{0}$

Substitute $e^{2j\pi r}$ for z:

$$H(f) = \frac{1}{5}\left[\frac{1}{e^{(2j\pi f)^4}} + \frac{1}{e^{(2j\pi f)^3}} + \frac{1}{e^{(2j\pi f)^2}} + \frac{1}{e^{(2j\pi f)^1}} + 1\right]$$

The effect of the *square* filter time-response is the appearance of ripples in the frequency domain plot of $|H(f)|$ (Fig. C.6). There are nulls in the frequency domain at $0.2 \times f_s$ and $0.4 \times f_s$, which is expected from the averaging process. The frequency axis extends from zero to the sampling frequency.

A filter with a smoother frequency domain response is the exponential smoothing filter, which is a simple example of the recursive filter structure shown in Fig. C.5b. To perform this filtration, at each sample we sum a fixed fraction (α) of the new sample value and (β) of the previous filtered value. In this example we provide also an overall gain K, sample interval Ts and the equivalent analog filter time constant Tc. When $K = 1$, $\alpha + \beta = 1$. In Mathcad this could be:

System gain: $K := 1.5$
Sample interval: $T_s := 0.1$
Time constant: $T_c := 0.5$
Gain coeffs: $\alpha := K\,T_s/(T_c + T_s)$ $\beta := T_c/(T_c + T_s)$
Calc. smoothing: $m := 1..N-1$ $U_m := \alpha V_m + \beta U_{m-1}$

Figure C.7, where one can guess that the shape of the response to a step input signal is a sampled exponential, very like a simple RC low-pass filter. When α is very small, each new measurement makes only small changes to the output value, and the system acts like a low-pass filter with a long time-constant. When $\alpha \approx 1$, the output rapidly follows the input, and smoothing is only that of a filter with a short time-constant. Again, programming in a math package or spreadsheet is easy. Here we only need to operate on the current input value and the previous output value, so calculation can begin from the second point of a dataset. Variation of the time-constant is achieved more simply through a change of the fixed value α, unlike the case of the sliding filter where both the dead-zone and the cells averaged need to change. The frequency-domain transfer function is shown in Fig. C.8.

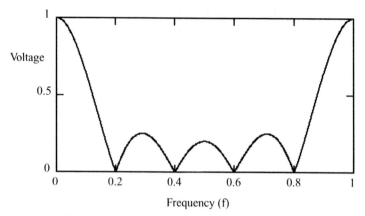

Figure C.6 Normalized frequency-domain response of the moving-average filter, showing ripples in the passband. Unity frequency represents the sampling rate.

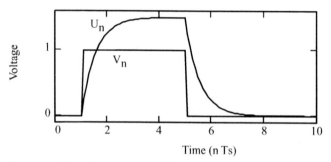

Figure C.7 Transient response of the recursive exponential filter.

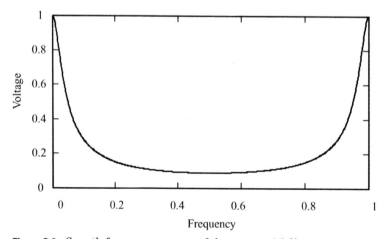

Figure C.8 Smooth frequency-response of the exponential filter.

A third type of nonrecursive filter is called the Savitzky-Golay technique, which is based on polynomial expansions of the functions represented by the data values. A five-point smoothing filter is given, as in a five-point sliding filter, by a sum of five data samples, but here they are weighted differently.

$$U_i = \frac{-3V_{i-4} + 12V_{i-3} + 17V_{i-2} + 12V_{i-1} - 3V_i}{35} \tag{C.6}$$

This has the effect of apodizing the time domain response. Figure C.9 shows an example. Small artifacts are seen at the discontinuous edges.

C.4   Differentiation in the Digital Domain

Occasionally we need to differentiate or high-pass filter a signal data-set. The simplistic way of taking differences between subsequent values and dividing by the time increment:

$$U_i = \frac{V_i - V_{i-1}}{t_i - t_{i-1}} \tag{C.7}$$

is often a poor approach. With noisy signals this can have the effect of greatly increasing the noise level. A much better approach is to use another of the nonrecursive Savitzky-Golay filters:

$$U_i = \frac{1V_{i-4} - 8V_{i-3} + 0V_{i-2} + 8V_{i-1} - 1V_i}{18} \tag{C.8}$$

By effectively fitting a parabola to three data values and determining the slope of the parabola, the differentiation process becomes much better behaved. Figures. C.10 and C.11 shows an example with the same step input as before, and also a ramp input.

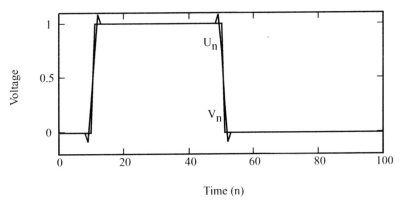

Figure C.9   Five-point Savitzky-Golay low-pass filter response.

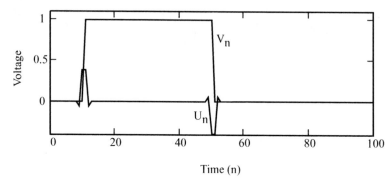

Figure   C.10 Five-point   Savitzky-Golay   high-pass   filter   response (differentiator).

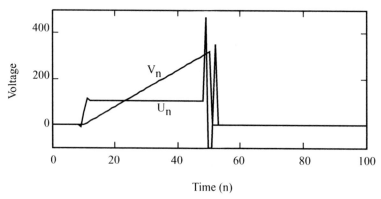

Figure C.11   Response of the five-point Savitzky-Golay differentiator to a ramp-input.

Whatever discrete-time filter type we use, we cannot escape the Nyquist sampling requirement. Signals should be restricted to a frequency low enough that two or more samples are taken per period, or aliasing will result. This can be seen in the frequency response of Fig. C.6, where only relative frequencies from 0 up to 0.5 are admissible. Above that, the response is symmetrically folded.

C.5   Averaging Multiple Measurements

When it comes to scanned measurements of sensor arrays, we have to make another decision on how to average. Let's say we want to perform a multi-channel measurement of the spectrum of an LED's emission (Fig. C.12). We have an eight-channel detector array viewing the dispersed spectrum. We measure each channel in turn, and then start the scan over again. We also want

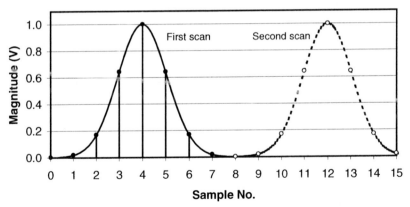

Figure C.12 Multiple-sensor arrays: Slow step and integrate or fast scan and average?

a spectrum recorded every second. We could integrate the signal from channel 0 for 125 ms, then move to channel 1 for 125 ms, and so on up to channel 7. Alternatively, we might sample channel 0 for just 1.25 ms, channel 1 for 1.25 ms and so on, afterwards repeating the scan 100 times and averaging the results of each channel. Clearly, the number of photons collected from each sensor should be the same in each case, if the light is more or less constant. So from the discussion of Chap. 3 the signal to noise of the two measurement strategies should be equal. The difference lies in the effective modulation frequencies of the detection process for each channel, and in the noise spectrum of the system.

Consider just one channel with the first strategy. It is integrated for a period of 125 ms, and then is ignored for 875 ms. In the frequency domain this gives a discrete spectrum of lines at 1 Hz intervals, modulated by an envelope function with a first null at 8 Hz (Fig. C.13). Hence much of the "modulation" energy is contained in the region from 0–8 Hz. When the second strategy is considered, the form of the discrete line spectrum is similar, but the lines are spaced at 100 Hz, with the first null of the envelope function at 800 Hz. Hence the energy is spread over a 100× wider region.

Note that we talk of modulation, although each channel of the LED is on continuously. For the signal processing it is immaterial whether the channel is illuminated for one-tenth of the time, or it is on continuously and we just measure it for one-tenth of the time. The advantages of modulating at about 800 Hz compared with 8 Hz are just the same as those considered in choosing the source modulation frequency for synchronous detection. It is generally better to avoid the frequencies where interfering noise sources are strongest. Hence scanning is closely related to modulation, and there are often great benefits in scanning fast and often, rather than slowly and less often.

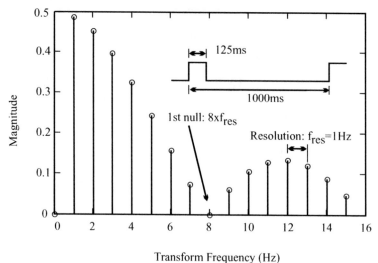

Figure C.13   Periodic sampling produces a discrete frequency-domain response consisting of lines separated by 1/total-scan-time, modulated by a sinc-function with a first null given by the ratio of total-scan-time to sample time (8 Hz).

If all eight channels are of equal interest, an even better strategy is to have eight complete measurement channels, so that each detector's output is used all the time. This should give eight times the number of collected photons and a S/N improvement of $\sqrt{8}$. This throughput advantage is often an argument to use a diode-array spectrometer instead of a scanned-slit spectrometer. If the whole spectrum is needed, the diode array device should give improved S/N. If only small regions of the total are of interest, however, then a scanned device set up to scan just that region may offer superior performance. This is the same criterion mentioned in Chap. 11 when considering use of an FFT or a conventional (slow) Fourier transform.

Digital signal processing is a large and complex subject, which has been well-treated in dozens of excellent books. As usual, the choice of book is a very personal one. My favorites seem not to be aimed purely at signal processing, but at digital control. Some are given in the references, and the reader is referred to them to learn more about z-domain analysis and filter synthesis techniques. I hope you have had a taste of the great possibilites of digital techniques, a glimpse at the elegant algebra of discrete time systems, and a warning about the dangers of under-sampled analog signals. In the same way that radio designers are abandoning analog filters and heterodyne techniques and reducing the receiver to an aerial, ADC and DSP chip, so too are optical measurement system designers leaving analog filters and lock-ins for DSP chips and software.

The Weighing Design Problem

Weighing designs look at how to optimally combine small objects whose weights are to be determined into larger collections placed simultaneously onto a beam- or spring-balance. With an analytical treatment of the summed measurements, the individual weights can be estimated with smaller relative errors. The techniques are equally applicable to measuring a "collection" of summed low intensities, whose magnitudes can be estimated with improved signal-to-noise ratio. This treatment is taken directly from Harwit and Sloane (1979).

Suppose we need to weigh four objects, with true but unknown weights ψ_1, ψ_2, ψ_3, ψ_4, with a beam-balance that shows a significant error e. The balance has been calibrated as well as possible, such that the average error is zero, and a large number of repeated measurements will actually deliver the object's true weight. If we weigh each object separately, we obtain four readings η_1, η_2, η_3, η_4, and four errors e_1, e_2, e_3, e_4. The readings must be taken as the best estimate of the true weights. The expectation value of the errors themselves is zero ($E\{e\} = 0$), while the expectation value of the squares of the errors ($E\{e^2\}$), is just σ^2, the variance of the measurements, or the mean square error in the weight estimates.

Now suppose that instead we weigh groupings of objects on a beam-balance as follows:

$$\eta_1 = \psi_1 + \psi_2 + \psi_3 + \psi_4 + e_1$$
$$\eta_2 = \psi_1 - \psi_2 + \psi_3 - \psi_4 + e_2$$
$$\eta_3 = \psi_1 + \psi_2 - \psi_3 - \psi_4 + e_3$$
$$\eta_4 = \psi_1 - \psi_2 - \psi_3 + \psi_4 + e_4$$

Here a plus sign signifies that the weight is placed on the left balance pan, a negative sign the right balance pan. The matrix of coefficients for the choice of weights is given by $H4$:

$$H4 = \begin{bmatrix} 1 & 1 & 1 & 1 \\ 1 & -1 & 1 & -1 \\ 1 & 1 & -1 & -1 \\ 1 & -1 & -1 & 1 \end{bmatrix} \text{ and its inverse: } H4^{-1} = \frac{1}{4}\begin{bmatrix} 1 & 1 & 1 & 1 \\ 1 & -1 & 1 & -1 \\ 1 & 1 & -1 & -1 \\ 1 & -1 & -1 & 1 \end{bmatrix}$$

The best estimates of the ψ may be obtained by multiplying the weight measurements by the inverse of $H4$:

$$(\psi_1) = \frac{1}{4}(\eta_1 + \eta_2 + \eta_3 + \eta_4)$$
$$= \psi_1 + \frac{1}{4}(e_1 + e_2 + e_3 + e_4)$$
$$(\psi_2) = \frac{1}{4}(\eta_1 - \eta_2 + \eta_3 - \eta_4)$$
$$= \psi_2 + \frac{1}{4}(e_1 - e_2 + e_3 - e_4)$$
$$(\psi_3) = \frac{1}{4}(\eta_1 + \eta_2 - \eta_3 - \eta_4)$$
$$= \psi_3 + \frac{1}{4}(e_1 + e_2 - e_3 - e_4)$$
$$(\psi_4) = \frac{1}{4}(\eta_1 - \eta_2 - \eta_3 + \eta_4)$$
$$= \psi_4 + \frac{1}{4}(e_1 - e_2 - e_3 + e_4)$$

The mean square error is therefore $E\{\frac{1}{16}(e_1 + e_2 + e_3 + e_4)\} = \frac{1}{4}\sigma^2$, or a factor four smaller than with individual object weighings. Note that this is the case where positive and negative measurements can be made. It corresponds to a referenced optical measurement in which a beam-splitter is used with two detector channels whose difference can take on positive and negative values.

The more common measurement of low level optical intensities using a single detector is more related to a weighing operation on a spring balance, where only positive measurements are possible. In this case the advantage of combining weights is reduced. For example, choosing

$$\eta_1 = \psi_2 + \psi_3 + \psi_4 + e_1$$
$$\eta_2 = \psi_1 + \psi_2 + e_2$$
$$\eta_3 = \psi_1 + \psi_3 + e_3$$
$$\eta_4 = \psi_1 + \psi_4 + e_4$$

from the coefficient matrix:

$$S4 = \begin{bmatrix} 0 & 1 & 1 & 1 \\ 1 & 1 & 0 & 0 \\ 1 & 0 & 1 & 0 \\ 1 & 0 & 0 & 1 \end{bmatrix} \text{ and its inverse: } S4^{-1} = \frac{1}{3}\begin{bmatrix} -1 & 1 & 1 & 1 \\ 1 & 2 & -1 & -1 \\ 1 & -1 & 2 & -1 \\ 1 & -1 & -1 & 2 \end{bmatrix}$$

In this case we can estimate the true weights from $S4^{-1}$:

$$(\psi_1) = \tfrac{1}{3}(-\eta_1 + \eta_2 + \eta_3 + \eta_4)$$
$$(\psi_2) = \tfrac{1}{3}(\eta_1 + 2\eta_2 - \eta_3 - \eta_4)$$
$$(\psi_3) = \tfrac{1}{3}(\eta_1 - \eta_2 + 2\eta_3 - \eta_4)$$
$$(\psi_4) = \tfrac{1}{3}(\eta_1 - \eta_2 - \eta_3 + 2\eta_4)$$

The optimum ways to choose objects for weighing on a beam balance are described by Hadamard matrices (e.g. $H4$ above). With N measurements, the mean square error is also reduced by N and the signal-to-noise increased by $\sqrt{N}$. The optimum ways to choose objects for weighing on a spring-balance, or equivalently single detector optical measurements, are described by the so-called S-matrices (e.g. S4 above). With N measurements, the mean square error is reduced by $(N + 1)^2/4N$ and the signal-to-noise increased by $(N + 1)/2\sqrt{N}$. For the small numbers of channels used in these examples, the gains in S/N are fairly modest. However, as N increases to the hundreds or thousands of resolved wavelengths typical of optical spectroscopy, or if the last decibel of S/N performance is needed from a low-light measurement, then these weighing-problem techniques are easily implemented and will provide the advantage.

References

Companies:

Advanced Photonix Inc., 1240 Avenida Acaso, Camarillo CA 93012. *www.advancedphotonix.com* (photodetectors, including large-area avalanche types).

Advanced Technology Coatings Ltd, Unit 1, Drakes Court, Eagle Road, Langage Business Park, Plympton, Plymouth, Devon PL7 5JY. *www.advanced-technology-coatings.co.uk*.

Alps Electric Co. Ltd, *www.opt.alps.com* (lenses for reduced-area, low capacitance photodiodes).

Centronic Ltd. Centronic House, King Henry's Drive, New Addington, Surrey, UK. Tel +44 (0)1689 808000. *www.centronic.co.uk* (high performance silicon photodiodes).

Comar Instruments Ltd., 70 Hartington Grove, Cambridge CB1 4UH, U.K. 01223-245470 (wide range of optical components).

Hamamatsu Corporation, 360 Foothill Road, P.O. Box 6910, Bridgewater NJ 08807-0910. *www.hamamatsu.co.uk* (wide range of photodetectors in silicon, and other semiconductors).

Jobin Yvon Horiba. *www.jyinc.com* (components and instruments for spectroscopy and industrial testing).

Infineon Technologies AG: www.infineon.com. Photodetectors are now handled by Osram Opto Semiconductors: *www.osram.convergy.de/search/product_class.asp*.

Laser Components (UK) Ltd., 4 Gloucester Avenue, Chelmsford, Essex CM2 9LD. *www.lasercomponents.co.uk*.

Mathsoft Inc. 101 Main Street, Cambridge MA02142 www.mathsoft.com. Mathcad® is a registered trademark of Mathsoft Engineering and Education, Inc.

MMG Neosid Ltd. Magnetic materials, RM-series pot-cores and toroids: *www.mmg-neosid.com*, and *www.mmg-sailcrest.co.uk*.

Melles Griot, 16542 Millikan Avenue, Irvine, CA 92606. *www.mellesgriot.com* (superb catalogue of optical components, an education in itself).

STEAG MicroParts GmbH., *www.steag-microparts.de* (compact spectrometer modules fabricated in LIGA technology).

Murata. *www.murata.com* (piezoelectric resonators for timing functions).

National Instruments, 11500 N. Mopac Expwy, Austin TX 78759-3504. *www.ni.com*.

Newport Corporation. *www.newport.com* (huge selection of optical elements and mechanical micro-manipulation hardware).

Ocean Optics Inc. 380 Main Street, Dunedin FL 34698. *www.OceanOptics.com* (Range of compact spectrometers).

Opto Diode Corp., 750 Mitchell Road, Newbury Park, CA 91320. *www.optodiode.com* (high-power LEDs).

OSI Fibercomm Inc. and UDT Sensors, Inc., 12525 Chadron Avenue, Hawthorne CA 90250. www.osifibercomm.com. *www.udt.com* (low capacitance, high-speed photodiodes).

Parallax Corporation. *www.parallax.com*. Basic Stamp® is a registered trademark. (easy to use Basic-programmable micro-controllers).

Philips. *www.semiconductors.philips.com* (specialist discrete transistors).

Rohm Electronics (U.K.) Ltd., Whitehall Avenue, Kingston, Milton Keynes, MK10 0AD, U.K. Tel: +44(1)908-282-666. *www.rohm.com*.

Schott Glass. *www.schott.com* (all things glass, including wide range of color filters).

Sumita Optical Glass Inc.: *www.sumita-opt.co.jp* (specialist glasses including fluorescent types).

TAOS *www.taosinc.com* (integrated detector/amplifiers, including self-clocked arrays).

Toshiba. *www.doc-semicon.toshiba.co.jp* (specialist discrete transistors).

Zeiss. *www.zeiss.com* (MMS UV compact spectrometer modules).

Semiconductors, Application Notes and Databooks:

Burr-Brown, *Applications Handbook, Application Bulletin* AB-075, 1994; "Photodiode Monitoring with Op Amps," *Application Bulletin* AB-061 "OPT201 Photodiode-amplifier rejects ambient light." 1993. *www.ti.com*.

Harris Corporation, *Linear ICs for Commercial Applications*, Harris Corporation, Melbourne, *www.harris.com*.

Hewlett Packard, "High Speed Fiber Optic Link Design with Discrete Components," Application note, 1022,1985.

Linear Technology, *Linear Applications Handbook*, 1990. *www.linear-tech.com*.

Maxim Inc. *New Releases Handbook*, Volume IV, *www.maxim-ic.com* (FET analog switches).

National Semiconductor Corporation "Linear Applications Databook," Many editions, e.g. 1986, also online at: *www.national.com*.

John Maxwell, *Low Noise FET Amplifiers*, March, 1977; *Transistor Databook*, 1982 (Circuits for coupling FETs to opamps).

Temic Inc. *Small-Signal Discrete Products*, Siliconix Inc.,1997. *www.temic.com*.

Zetex, Inc., Zetex, Applications Handbook 1996. Application Note: AN3 "Infra-red Remote Control and Data Transmission," *www.zetex.com* (high performance discrete bipolar transistors and FETs).

Books and Articles:

Allen, M.G., et al., "Ultrasensitive Dual-Beam Absorption and Gain Spectroscopy," *Applied Optics*, 34:3240–3249, 1995.

Ambrozy, A., *Electronic Noise*, McGraw-Hill, New York, 1982.

Barone, F., et al., "High Accuracy Digital Temperature Control for a Laser Diode," *Review of Scientific Instruments* 66:4051–4054, 1995.

Bass, M., Ed., *Handbook of Optics, Encyclopedia of Optical Techniques*, Volume 2, "Devices, Measurement and Properties," 2d ed., Optical Society of America and McGraw-Hill, New York, 1995.

Baumeister, P., "Narrow-Bandpass Filters Meet Performance Criteria," *Laser Focus World*, February: 103–110, 1989.

Baxendall, P.J., "Noise in Transistor circuits," *Wireless World* 388–392 November 1968.

Beauchamp, K.G., *Walsh Functions and Their Applications*, Academic Press, New York, 1975.

Berlin, H.M., *Design of Active Filters, with Experiments*, Howard Sams, Indianapolis, 1977.

Beslich, P.W., "Walsh Function Generators for Minimum Orthogonality Error," *IEEE Transactions on Electromagnetic Compatibility EMC-15*,1973.

Boys, C.V., *Soap bubbles*, Dover Publications Inc., New York, 1959.

Carr-Brion, K.G., and Dowdeswell, R.M., "Non-Fouling Optical Windows for On-Line Analysis", WWT, p14–16, October 1996.

Fecht, I., and Johnson, M., "Non-Contact, Scattering-Independent Water Absorption Measurement Using a Falling Stream and Integrating Sphere," *Measurement Science and Technology*, 10:612–618, 1999.

Felgett, P., *The Theory of Infrared Sensitivities and Its Application to Investigations of Stellar Radiation in the Near Infrared*, Ph.D. thesis, Cambridge University, Cambridge, 1951.

Felgett, P., "Conclusions on Multiplex Methods," *Journal de Physique*, Colloque C2, 28:165–171, 1967.

Forsberg, G.S., "Optical Receiver with Optical Feedback," *Electronics Letters* 23:478–480, 1987.

Fowler, B., Balicki, J., How, D., Godfrey, M., "Low FPN High Gain Capacitive Transimpedance Amplifier for Low Noise CMOS Image Sensors". *SPIE Conference Sensors, Cameras, Systems for Scientific/Industrial Applications III*. San Jose, January 2001.

Golten, J., and Verwer, A., *Control System Design and Simulation*, McGraw-Hill, New York, 1997.

Goodman, J.W., *Statistical Optics*, Wiley International, New York, 1985.

Göpel, W., Hesse, J., Zemel, J.N., Ed., *Sensors, A Comprehensive Survey*, Volume 6, *Optical Sensors*. VCH Weinheim 1992.

Graeme, J., *Photodiode Amplifiers*, McGraw-Hill, New York, 1995.

Grocock, J.A., "Design with Paralleled Transistors for Use with Low Source Resistance," *Wireless World*, March 1975.

Harmuth, H., "Grundzüge einer Filtertheorie für die Mäanderfunktionen $A_n(\Theta)$,"*AEU* 18:544–554, 1964.

Harwit, M., and Sloane, N.J.A., *Hadamard Transform Optics,* Academic Press, New York, 1979.

Hickman, I., "Reflections on Opto-Electronics," *Electronics World and Wireless World,* November: 970–974, 1995.

Hickman, I., *Analog Circuits Cookbook,* Newnes-Butterworth-Heinemann. Woburn Ma., 2001.

Hickman, I., "Pseudo-Random Bits," *Electronics World,* July: 500–503, 2001.

Hobbs, P.C.D., "Ultrasensitive Laser Measurements Without Tears," *Applied Optics,* 36:903–920, 1997.

Hobbs, P.C.D., *Building Electro-Optical Systems,* Wiley-Interscience, New York, 2000.

Hodgkinson, J., Johnson, M., and Dakin, J.P., "Photothermal Detection of Trace Optical Absorption in Water by Use of Visible-Light-Emitting Diodes," *Applied Optics,* 37:7320–7326, 1998.

Hodgkinson, J., Johnson, M., and Dakin, J.P., "Photothermal Detection of Trace Compounds in Water Using the Deflection of a Water Meniscus," *Measurement Science and Technology,* 9:1316–1323, 1998.

Houser, G.D., and Garmire, E., "Balanced Detection Technique to Measure Small Changes in Transmission," *Applied Optics,* 33:1059–1062, 1994.

Imasaka, T., Kamikubo, T., Kabawata, Y., Ishibashi, N., "Ultratrace Photometric Determination of Phosphate with a Solid-State Emitter as Light Source," *Analytica Chimica Acta,* 153:261–263, 1983.

Jackman, R.B., "Diamond Photodetector is Blind to Visible Light,", *Optics and Laser Europe,* September: 19–25, 1996.

Jenkins, K.W., *Teach Yourself Algebra for Electric Circuits,* McGraw-Hill, New York, 2002.

Johnson, M., and Jokerst, N.M., "Self-Detecting Light-Emitting Diode Optical Sensor," *Journal of Applied Physics,* 56:869–871, 1984.

Johnson, M., and Stäcker, D., "A Non-Fouling Optical Interface for Environmental Measurements," *Measurement Science and Technology,* 9:399–408, 1998.

Johnson, M., and Melbourne, P., "Photolytic Spectroscopic Quantification of Residual Chlorine in Potable Waters," *Analyst,* 121:1075–1078, 1996.

Johnson , M., "Spectroscopic Detection of Aqueous Contaminants Using In Situ Corona Reactions," *Analytical Chemistry,* 69:1279–1284, 1997.

Johnson, M., and Melbourne, P., Photolytic Spectroscopic Detection of Herbicides and BTEX in Water, In: Lampropoulos, G.A., and Lessard,R.A., *Applications of Photonic Technology,* Second Edition. Plenum Press, New York, 1997.

H. Kaplan, H., "Black Coatings Are Critical in Optical Design," *Photonics Spectra,* January:48–49, 1997.

Kasper, B.L., McCormick, A.R., Burrus, C.A., Talman, J.R., "An Optical-Feedback Transimpedance Receiver for High Sensitivity and Wide Dynamic Range at Low Bit Rates." *IEEE Journal of Lightwave Technology,* 6:329–338, 1988.

Katz, P., *Digital Control Using Microprocessors,* Prentice/Hall International, London, 1981.

Krattinger, B., et al., "Laser-Based Refractive-Index Detection for Capillary Electrophoresis: Ray-Tracing Interference Theory," *Applied Optics,* 32:956–965, 1993.

Langton, C.H., "Orthogonal Functions—An Introduction to Walsh Functions" *Electronics and Wireless World,* 283, 1985.

Larson, A.P, Ahlberg, H., Folestad, S., "Semiconductor Laser-Induced Fluorescence Detection in Picoliter Volume Flow Cells," *Applied Optics,* 32:794–805,1993.

Lathi, B.P., *Linear systems and signals,* Oxford University Press, New York, 2002.

Liu, H., Dasgupta, P.K., Zheng, H.J., "High Performance Optical Absorbance Detectors Based on Low-Noise Switched Integrators," *Talanta* 40:1331–1338, 1993.

Liu, S., and Dasgupta, P.K., "Liquid Droplet. A Renewable Gas Sampling Interface," *Analytical Chemistry,* 67:2042–2049, 1995.

Longhurst, R.S., *Geometrical and Physical Optics,* Longman, London, 1968.

Marsh, D., "Op Amps Take the Next Step," *Electronic Design News Europe,* 22–33 September: 22–33, 2002.

Martens, H., and Næs, T., *Multivariate Calibration,* John Wiley & Sons, New York, 1993.

May, D., *Hickman's Analog and RF Circuits,* Elsevier, New York, 1998.

Mims, F.M., III, "Solar Radiometer with Light-Emitting Diodes as Spectrally Selective Detectors," *Optics and Photonics News,* Supplement, 11:2, 3–4, Feb. 2000.

Netzer, Y., "The Design of Low-Noise Amplifiers," *Proceedings of the IEEE,* 69:728–741, 1981.

Pease, R., "What's All This Transimpedance Amplifier Stuff, Anyhow?" *Electronic Design,* January 8:139–144, 2001.

Pease, R., "What's All This Femtoampere Stuff, Anyhow?" *Electronic Design* September 2, 1993.

Pease, R., "What's All This Teflon Stuff, Anyhow?" *Electronic Design*, Feb 14, 1991.

Philip, R.F., "High Gain High Bandwidth Amplification Using Two Op-amps," *Electronic Product Design*, September: 21–22, 1995.

Reider, G.A., Traar, K.P., Schmidt, A.J., "Thin Fluid Jet Stream of High Optical Quality," *Applied Optics*, 23:2856–2857, 1984.

Sheingold, D.H., Ed. "Transducer Interfacing Handbook," *Analog Devices*, 1981.

Skoog, D.A., Holler, F.J., Nieman, T.A., *Principles of Instrumental Analysis*, 5th ed., Saunders College Publishing, Philadelphia, 1998.

Stanley, W.D., Dougherty, G.R., Dougherty, R., *Digital Signal Processing*, Reston Publishing Company, Reston, VA 1984.

Steffes, M., "Embedded Gain Supercharges FET-Transimpedance Amplifier," *Electronic Design News*, May 22, 1997.

Steffes, M., "Here's an Easy Way to Test Wideband Transimpedance Amplifiers." *Electronic Design*, June:74–80, 1998.

Stout, D.F., *Handbook of Operational Amplifier Circuit Design*, McGraw-Hill, New York, 1976.

Sze, S.M., *Physics of Semiconductor Devices*, Wiley International Edition, New York, 1969.

Ulrich, R., and Rashleigh, S.C., "Beam-to-Fiber Coupling with Low Standing Wave Ratio," *Applied Optics*, 19:2453–2456, 1980.

Umbach, A., Trommer, D., and Unterbörsch, G., "Photodetectors for 40 Gb/s Applications and Beyond," *Europhotonics*, August/September: 52–53, 2001.

Verma, K.K., Jain, A., Townshend, A., "Determination of Free and Combined Residual Chhlorine by Flow-Injection Spectrophotometry," *Analytica Chimica Acta*, 261:233–240, 1992.

Woodward, W.S., "Self-heated Transistor Thermostats Individual Components," *Electronic Design*, February 9:136–137, 1998.

Yates, F., "Complex Instruments," *Journal of the Royal Statistical Society*, 2:181–247, 1935.

Yoldas, B.E., and Partlow, D.P., "Wide Spectrum Antireflective Coating for Fused Silica and Other Glasses," *Applied Optics*, 23:1418–1424, 1984.

Index

ABOUT THE AUTHOR

Mark Johnson, Ph.D., is an independent consultant in opto-electronics and measurement innovation. He is a visiting professor at Salford University in England and St. Etienne University in France and has managed corporate research teams in the United Kingdom, the United States, and Germany. Dr. Johnson resides in Cheshire, England.